BIOCHEMISTRY RESEARCH TRENDS

SPHINGOLIPIDS

BIOLOGY, SYNTHESIS AND FUNCTIONS

BIOCHEMISTRY RESEARCH TRENDS

Additional books in this series can be found on Nova's website
under the Series tab.

Additional e-books in this series can be found on Nova's website
under the e-book tab.

Library of Congress Cataloging-in-Publication Data

Sphingolipids : biology, synthesis, and functions / editor, Angel Catala (Facultad de Ciencias Exactas, Universidad Nacional de La Plata, Sucursal, La Plata, Argentina).
pages cm. -- (Biochemistry research trends)
Includes index.
ISBN 978-1-63483-019-5 (hardcover)
1. Sphingolipids. I. Catala, Angel.
QP752.S6S6864 2015
571.6--dc23
2015031083

Published by Nova Science Publishers, Inc. † New York

BIOCHEMISTRY RESEARCH TRENDS

SPHINGOLIPIDS

BIOLOGY, SYNTHESIS AND FUNCTIONS

ANGEL CATALÁ
EDITOR

nova publishers
New York

CONTENTS

PREFACE

This book presents an overview of sphingolipids: biology, synthesis and functions. The topics analyzed in this book, cover a broad spectrum of functions played by sphingolipids including: the role of sphingomyelin in the regulation of membrane physical state; tricyclic antidepressants as modulators of sphingolipid turnover and cell sensitivity to insulin action; sphingolipid diversity, biosynthesis and regulation; metabolism of glycolipids at the cell's surface: modulation and functional implications; the specific role of sphingomyelins in cell physiology; the role of sphingosine-based lipids using sphingolipidomics; sphingolipids and ceramides of aqueous humor and trabecular meshwork; the effect of membrane microdomains containing sphingolipids on lipid peroxidation; and roles of sphingolipids in skeletal development and homeostasis.

Chapter 1 – Cell membranes are complex structures made of hundreds of different types of molecules. Their physical state is function of lipid and protein composition, temperature and extracellular conditions. The physical state of a cell membrane regulates the activity of all embedded proteins, as well as the expression of stress genes to maintain homeostasis. Sphingolipids are a major component of cell membranes and they have key roles as signaling molecules in a wide range of biological processes. The morphology of sphingolipids can influence the lipid phase order and the overall fluidity of membranes. Sphingolipids are usually found associated with sterols to form dynamic lateral membrane "rafts" [1]. These lipid domains are responsible for the interaction with membrane proteins and thus regulate their biogenesis and activity [1, 2]. Additionally, specific signaling proteins are capable to discriminate among liquid-ordered (L_o) and liquid-disordered (L_d) compartments [3]. The raft structure is not only dependent on the lipid composition but temperature or pressure can influence the biophysical state and therefore even mild variations in temperature may well result in a profoundly altered solubility and consequently redistribution of proteins in membranes. These changes may affect the activity of certain membrane proteins [4, 5]. In the last years, a membrane sensor model has been proposed [6] which considers the role of lipid membranes as stress sensor and signaling mechanisms in all types of cells. It is then possible to rationalize the ability of some lipophilic drugs to modulate the activity of membrane proteins, altering their lateral localization within a membrane [7, 8]. Reasonably, the overall changes in the membrane fluidity, remodeling of specific domains, and/or changes in composition of specific lipid/protein interactions, are potentially able to provide stimuli for the activation of heat shock genes.

In this review, the role of sphingomyelin and of sphingomyelin/cholesterol ratio in regulating the cell stress response is considered along with the possibility to exploit small molecules to interact with lipid membranes, and consequently perturb the proteins embedded in membranes.

Chapter 2 – Because of their effect on the reuptake of norepinephrine or serotonin tricyclic antidepressants are used to treat the depressive disorders. Such antidepressants as amitriptyline, desipramine, imipramine, nortriptyline, doxepin, clomipramine, and protriptyline are the more commonly used. Findings from cellular and animal studies demonstrated that antidepressant drugs may regulate ceramide levels and decrease acid sphingomyelinase (aSMase) activity. Such tricyclic antidepressants as desipramine and imipramine cause via enzyme proteolytic degradation a detachment of the aSMase from the inner membrane and subsequent inactivation. These drugs do not inhibit aSMase directly, but may act as functional inhibitors of SMases and thus produce the anti-apoptotic, proliferative and anti-inflammatory effects. Moreover, the aSMase inhibitor, amitriptyline, attenuated obesity and obesity-associated renal injury, induced by the high-fat-diet treatment. Inhibition of aSMase by desipramine significantly reduced triacylglycerol synthesis and content and increased the (dihidro)ceramide synthesis in the HepG2 cells, cultured in the presence of exogenous palmitic acid. These results provide direct evidence that aSMase supplies the accumulated palmitic acid mainly to sphingolipid synthesis. In contrast, the overexpression of aSMase in the diabetic db/db mice is associated with glucose and lipid metabolism improvement. aSMase inhibition with imipramine decreases glucose uptake by liver cells and glycogen and lipid deposition in hepatocytes. However, inhibition of aSMase by imipramine partly improves the hepatocyte sensitivity to insulin action in the aged rats. Treatment of "old" insulin-resistant cells with imipramine reduces the elevated aSMase activity and ceramide synthesis and increases insulin-induced glucose uptake and glycogen synthesis. Although the functional inhibitors of aSMase are widely used in cellular and animal studies to improve glucose and lipid homeostasis, the therapeutic consequences of functional inhibition of aSMase are yet poorly understood. However, most of the available tricyclic antidepressants are licensed for medical use in humans, as being minimally toxic and therefore may be used for treatment of disease states associated with increased activity of aSMase. Recently functional inhibitors of aSMase have been shown to reduce the severity of different animal models of human diseases, including cystic fibrosis, Wilson disease, acute lung injury and hepatic ischaemia-reperfusion injury. This chapter will be focused on the role of tricyclic antidepressants in the regulation of sphingolipids turnover in the cells at different pathological conditions associated with enhanced activity of aSMase. Much attention will be given to the modulation of cell sensitivity to insulin action by functional inhibitors of aSMase activity.

Chapter 3 – Sphingolipids constitute a ubiquitous class of lipids present in all eukaryotes as well as in several prokaryotes and even some viruses. Their structure is peculiar among other classes of lipids with a backbone made of a sphingoid base bearing an N-linked fatty acid. This structure, known as ceramide, is further modified by means of polar head substitutions. Head substitutions, fatty acids of different lengths and sphingoid bases variants generate a great assortment of sphingolipid isoforms (sphingolipidome), which is highly diversified between organisms, kingdoms and cell types. Sphingolipids are essential for organism development and cell growth since complex sphingolipids are major components of biological membranes and involved in membrane domain organization and receptor signaling. Sphingoid bases are instead involved in cell signaling, cell growth and stress responses. Several key enzymes of the biosynthetic pathway are highly conserved and finely regulated to generate key "structural orthologs" that seem to be shared among organisms. In this chapter, the authors describe step by step the sphingolipid biosynthetic pathway, and via a continuous parallel between plant, yeast and animal, they will review the sphingolipid functions in different organisms and cells.

Chapter 4 – Glycosphingolipids (GSLs) are amphipathic molecules consisting of a ceramide lipid moiety linked to a glycan chain of variable length and structure. Among these are found the gangliosides, which are mono- or multi-sialosylated GSLs ubiquitously distributed on the external leaflet of vertebrate plasma membranes. Gangliosides have been implicated in many physiological processes, including growth, apoptosis, cell adhesion and endocytosis, as well as in pathological processes due to their capacity to function as receptors for several toxins, viruses and antibodies. Their *de novo* synthesis starts at the endoplasmic reticulum and continues by a combination of glycosyltransferase activities at the Golgi complex, followed by vesicular delivery to the plasma membrane. At the cell surface, gangliosides can undergo endocytosis, and once internalized they can be sorted in a cell-type-dependent manner into endosomes for recycling to the plasma membrane or to be catabolized at the lysosomal level. However, a number of enzymes for ganglioside anabolism and catabolism have recently been shown to associate with the plasma membrane, being able to exert their enzymatic activities on substrates residing in the same surface (cis-catalytic activity) and in that of neighboring cells (trans-catalytic activity). The action of these enzymatic activities may be responsible for structural changes of the plasma membrane gangliosides, thus modifying membrane organization, and consequently, different signaling processes. In this Chapter, the authors review early and recent advances concerning the glycolipid metabolism at the cell surface, as well as its possible biological implications, which represents a new paradigm in the cellular control of glycolipid expression.

Chapter 5 – Sphingomyelin (SM) is one of the most abundant components of cellular membranes in organisms ranging from mammals to protozoa accounting for 10 to 20% of total phospholipid content. It is preferentially concentrated at the plasma membrane outer leaflet and its high packing density and affinity for sterols provide a rigid barrier to the extracellular environment and play a role in the formation of lipid rafts. SM consists of an sphingoid backbone linked to a fatty acid and to phosphocholine moiety. When analyzing SM molecular composition up to four different sphingoid bases and more than 15 different fatty acids may be found, the combination of which account for more than hundred different SM molecular species.

A recent study showing the direct interaction between one specific SM molecular species (SMd18:1-18:0) and a protein involved in COPI-transport (p. 24) indicates that molecular species of a particular lipid class may be as important as the different protein isoforms. Therefore, it is critical to accurately determine the lipid composition when the function of a lipid class is studied. The rapid advances in mass spectrometry applied to lipidomics will definitively help to tailor the specific role that each of the different SM may have in cell physiology.

SM has been traditionally viewed as a simple structural building block of the plasma membrane. Currently, it is accepted that in addition to its structural functions, SM serves as source of important signaling molecules. However, the possibility that SM by itself could be important in regulating crucial cell events has been clearly underestimated despite the existing evidence. Thus, phorbol ester-induced differentiation of HL-60 human promyelocytic leukemia cells into macrophages is coupled to an increase in SM synthesis via Sphingomyelin Synthase (SMS). Conversely, SM hydrolysis induces proliferation of quiescent 3T3 fibroblasts. Similarly, quiescent satellite cells (skeletal muscle stem cells) contain higher SM levels than proliferating satellite cells where SM was barely detected. Furthermore, non-tumor cells contain higher SM levels than a wide range of tumor cell lines. Altogether, these studies show that SM content in cells in a non-proliferating state (i.e. quiescent, non-tumor, and differentiated cells) is higher than in cells in a proliferating state (activated, tumor and non-differentiated cells) indicating that SM levels may play a role in the control of cell cycle progression. In this context, it has been demonstrated that while SM synthesis plays an active role in cell adhesion, its degradation after SMase treatment looses attached cells, indicating that SM level is indeed important in cell adhesion regulation. The wide range of cell lines where this phenomenon occurs suggests that SM content regulation may be a general strategy used by cells to regulate their fate. Finally, studies using acid sphingomyelinase (ASM-) KO mice demonstrate that SM levels are critical for brain functioning. Thus, the increased SM levels found in ASM-KO neurons results in reduced spine number and size and, low levels of filamentous actin. Overall, it is clear that the role of SM in cell function goes beyond being a mere structural element or a source of signaling molecules.

Chapter 6 – Sphingolipids, a heterogeneous class of lipids, are key players as constituents of cell membranes, and as signalling molecules. Bioactive sphingolipids mediate many key cellular processes, including cell metabolism, growth, differentiation, stress and inflammatory responses, cell apoptosis and autophagy. Altered sphingolipids profiles have been frequently observed in various diseases, such as inflammatory and neurodegenerative diseases, cancer and sphingolipidoses. They are in particularly relevant in Gaucher and Fabry diseases.

The development of mass spectrometric methods and technologies has improved the comprehensive analysis of the total sphingolipid profile, and allows one to decipher their biological functions in health and disease. Based on different analytical approaches, sphingolipidomics has been used for the molecular characterization and quantification of sphingolipid classes and molecular species. This chapter describes the basis of modern mass spectrometry approaches to sphingolipidomics, and the application to the study of sphingolipids in living systems. Special emphasis is placed on disease states.

Chapter 7 – The authors present a review of recently analyzed sphingolipid and ceramide profiles of human control and primary open angle glaucoma (POAG) trabecular meshwork (TM) and aqueous humor (AH). The following three classes of lipids: sphingomyelin, sphingoid base, sphingoid base-1-phosphate have been collectively termed as sphingolipids here. The original published profiles were generated by extensive class specific mass spectrometry. Comparisons of respective profiles between control and POAG (TM or AH) showed a majority of sphingolipids and ceramides to be common and very few unique species associated with a given state (control or diseased state) between these two groups. In human TM, sphingomyelin (SM) species were higher in control than in POAG, while the opposite was observed in AH. In human controls, sphingoid base (SB) species and ceramides (Cer) were higher in both TM and AH. Sphingoid base-1-phosphates (SB1P) was found elevated in control AH but the levels remained the same for control and POAG in the TM.

In order to test for changes in the sphingolipid and ceramide profiles between normotensive and hypertensive states the authors repeated the analyses with DBA/2J mice (a mouse model of glaucoma, which undergoes spontaneous transition from normotensive to hypertensive state). They compared TM and AH from the normotensive (mice with an IOP of ≤ 17 mmHg) and the hypertensive state (with an IOP of > 18 mmHg). Since the normotensive mice are younger in age (3-6 months) than those in the hypertensive state (8-10 months in age) thus the aging changes are not completely ruled out from these studies. The sphingolipid and ceramide profiles of hypertensive state DBA/2J showed similarity with human POAG with respect to both the TM and AH. The male and female DBA/2J mice showed variation in their lipid composition of TM and AH, but the overall trend of each of the four lipid species showed similar changes. SM and SB species were higher in the normotensive state for both AH and TM. SB1P and ceramides were lower in normotensive AH and TM. The comparison of human control, POAG and mice normotensive, hypertensive states gives insight into the changes that occur in the hypertensive state without any drug treatment for lowering intraocular pressure. The authors also present preliminary imaging mass spectrometry (IMS) data for sphingomyelin. The authors' current IMS results are consistent with extractive mass spectrometry findings.

Chapter 8 – Sphingolipids are an integral part of lipid membranes, primarily the plasma membrane of cells. It was thought that these macromolecules were inert structures that were present in cell membranes to maintain cell structure and to serve as adhesion sites for proteins. However, this view has changed throughout the years as recent studies show that sphingolipids and their metabolites have downstream effects on cell signal-transduction pathways, which regulate cell differentiation, proliferation and apoptosis. Recent data also give importance to sphingolipids as regulatory molecules in skeletal tissues. Sphingolipid metabolism is critical for fundamental cellular events and this chapter will discuss current data showing the various roles of some key sphingolipids and their bioactive metabolites in skeletal development and tissue homeostasis.

Chapter 9 – Sphingomyelin (SM) is a major phospholipid in cell membranes and lipoproteins of all animals. Its concentration fluctuates considerably between different cells, and even among membranes of the same cell. An interesting characteristic of SM distribution is that is similar the distribution of free (unesterified) cholesterol (FC). For instance, both these lipids are concentrated in the plasma membrane of the mammalian cells, and are important constituents of the membrane rafts which are involved in a broad diversity of cellular functions. Lipid rafts are microdomains of the plasma membrane that contain high levels of cholesterol and sphingolipids. In terms of lipids, the heterogeneous membrane is believed to consist of a mixture of a dispersed ''lipid raft" phase, enriched in cholesterol, raft-associated proteins, and saturated lipids (such as sphingolipids), and the ''non-raft" matrix phase enriched with phospholipids containing polyunsaturated fatty acids (PUFAs) and vitamin E. Phospholipids are essential components of cell membranes that maintain structure and control the function of cells. Lipid peroxidation is one of the major outcomes of free radical-mediated injury to tissue. Peroxidation of fatty acyl groups occurs mostly in membrane phospholipids located in the ''non-raft" matrix phase. The dispersed ''lipid raft" phase, enriched in cholesterol, raft-associated proteins, and sphingolipids is not affected by lipid peroxidation because sphingolipids contain only saturated fatty acids in its molecule, Peroxidation of lipids can greatly alter the physicochemical properties of membrane lipid bilayers, resulting in severe cellular dysfunction. In addition, a variety of lipid by products is produced as a consequence of peroxidation. Lipids containing polyunsaturated fatty acids are subject to free radical–initiated oxidation and can contribute in chain reactions that amplify damage to biomolecules.

In: Sphingolipids
Editor: Angel Catalá
ISBN: 978-1-63483-019-5

Chapter 1

THE ROLE OF SPHINGOMYELIN IN THE REGULATION OF MEMBRANE PHYSICAL STATE

Stefano Piotto[1,*], Lucia Sessa[1] and Simona Concilio[2]
[1]Department of Pharmacy, University of Salerno, Fisciano (SA), Italy
[2]Department of Industrial Engineering,
University of Salerno, Fisciano (SA), Italy

ABSTRACT

Cell membranes are complex structures made of hundreds of different types of molecules. Their physical state is function of lipid and protein composition, temperature and extracellular conditions. The physical state of a cell membrane regulates the activity of all embedded proteins, as well as the expression of stress genes to maintain homeostasis. Sphingolipids are a major component of cell membranes and they have key roles as signaling molecules in a wide range of biological processes. The morphology of sphingolipids can influence the lipid phase order and the overall fluidity of membranes. Sphingolipids are usually found associated with sterols to form dynamic lateral membrane "rafts" [1]. These lipid domains are responsible for the interaction with membrane proteins and thus regulate their biogenesis and activity [1, 2]. Additionally, specific signaling proteins are capable to discriminate among liquid-ordered (L_o) and liquid-disordered (L_d) compartments [3]. The raft structure is not only dependent on the lipid composition but temperature or pressure can influence the biophysical state and therefore even mild variations in temperature may well result in a profoundly altered solubility and consequently redistribution of proteins in membranes. These changes may affect the activity of certain membrane proteins [4, 5]. In the last years, a membrane sensor model has been proposed [6] which considers the role of lipid membranes as stress sensor and signaling mechanisms in all types of cells. It is then possible to rationalize the ability of some lipophilic drugs to modulate the activity of membrane proteins, altering their lateral localization within a membrane [7, 8]. Reasonably, the overall changes in the membrane fluidity, remodeling of specific domains, and/or changes in composition of specific lipid/protein interactions, are potentially able to provide stimuli for the activation of heat shock genes.

* Email: piotto@unisa.it

In this review, the role of sphingomyelin and of sphingomyelin/cholesterol ratio in regulating the cell stress response is considered along with the possibility to exploit small molecules to interact with lipid membranes, and consequently perturb the proteins embedded in membranes.

1. INTRODUCTION

1.1. Membrane Lipid Composition

In recent years, the concept of a biological membrane evolved from the concept of a simple physical barrier with extremely low water permeability into a highly sophisticated system capable to control protein activity, intracellular signaling, and stress response. Membrane lipids are usually classified into three main groups: ceramide-based sphingolipids, glycerol-based lipids, and sterols. There are two categories of glycerol-based lipids: phospholipids and glycosylglycerides. In phospholipids, the *sn*-1 and *sn*-2 positions are esterified to a fatty acid, the *sn*-3 position is esterified to a phosphate group and then esterified to a polar head group. Though the fatty acid portion is critical to determine the overall properties, the polar head group is usually used for classification. Glycosylglycerides are the most abundant membrane glycerolipids. In glycosylglycerides the *sn*-3 position of the glycerol backbone is esterified to a glycosyl group. Sterols, with cholesterol being the most important member of the group, contain a hydroxyl moiety interacting with the head groups of sphingomyelins and phospholipids. The hydroxyl group remains usually protected from water by the so called *umbrella effect.* The bulky steroid portion interacts strongly with saturated lipid acyl chains. This interaction regulates membrane packing and fluidity, and the formation of microdomains. The presence of a sphingoid-base backbone defines the sphingolipid class. The chemical physical properties are largely superimposable to those of phospholipids. In eukaria sphingolipids are usually found associated with cholesterol molecules, and therefore they are particularly abundant in raft microdomains.

1.2. Sphingomyelin and Ceramide

Sphingolipids are involved in diverse cell functions and their metabolism is controlled by numerous enzymes. Ceramide (Cer) occupies a central position in the sphingolipid network, being involved in cell death, differentiation, senescence, autophagy, and migration [9, 10]. Ceramide is produced with 3 pathways: *de novo* synthesis, salvage pathway, and sphingomyelin (SM) cycle *via* sphingomyelinase (SMase) (Figure 1). Furthermore, SM can serve as substrate for the synthesis of other sphingolipids. Ceramide has a well-known role in regulating cell death and survival. Sphingomyelins have a role in cell migration, apoptosis, autophagy, and cell proliferation and they are also involved in human disorders such as cancer.

In the salvage pathway the sphingosine (Sph), which is metabolized from ceramide or S1P, is acylated by ceramide synthase (CerS) to give ceramide [11]. There are six isoforms of CerS (CerS1-6) with a different specificity for each carbon chain length of fatty acyl CoAs. They can produce distinct ceramides having different acyl-chain lengths. Moreover, CerS can act as dihydroceramide synthase. This means that CerS is essential for ceramide generation in both *de novo* synthesis and salvage pathway [12]. In the proposed "SM cycle," ceramide is generated *via* hydrolysis of SM by the activation of SMases [13]. Conversely, SM can be produced from ceramide *via* Sphingomyelin synthase (SMS). SMS catalyzes the transfer of the phosphorylcholine group from phosphatidylcholine (PC) to Cer. This leads to the formation of SM and Diacylglycerol (DAG) [14].

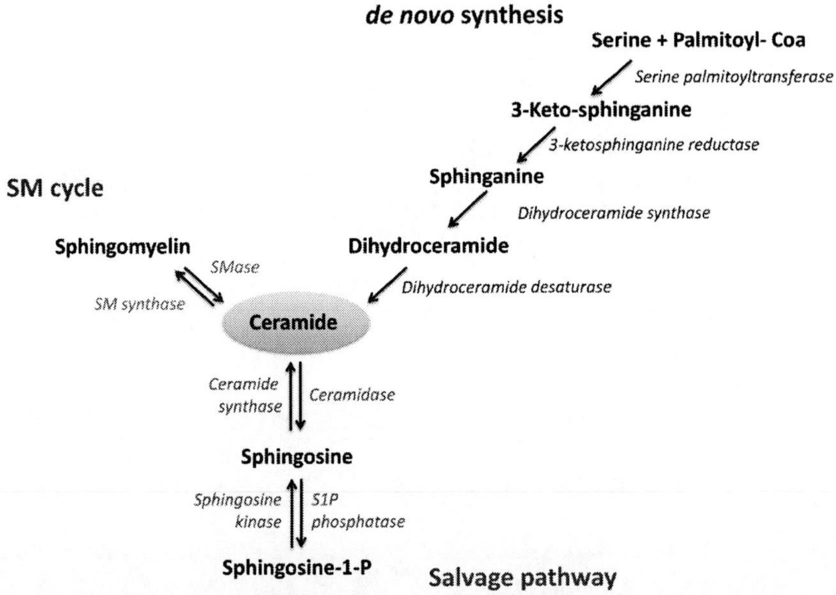

Figure 1. Biosynthesis of sphingolipids (enzymes in italics).

In 2004, the SMS1 gene, named SGMS1, was identified from the human cDNA library [15]. The sequence analysis revealed several subsequences in common with lipid phosphate phosphatase (LPP) and numerous transmembrane domains. Later, SMS genes in the human genome were identified as SGMS1 (SMS1), SGMS2 (SMS2), and SAMD8 (SMSr). The proteins of human SMS1, SMS2, and SMSr consist of 413, 365, and 414 amino acids (aa), respectively. SMS1 protein is highly conserved through species.

Hydrophobicity analysis postulates that both SMS1 and SMS2 contain six transmembrane α-helices connected by extramembrane loops domains. They shared high sequence similarity, however in the SMS2 the SAM domain is missing (Figure 2).

In SMS1 the *N*-terminal (SAM domain) and the *C*-terminal face the cytosol [16]. The SAM domain may play a role in cellular functions, such as development, signal transduction, and transcriptional regulation through protein–protein or protein–lipid interaction [17-20]. Nevertheless the proteins interacting with SAM domain of SMS1 have not yet identified, and SAM domain function remains unclear.

Point mutations in SMS1 and SMS2 sequences inactivate SM synthesis without changing cellular distribution. *C*-terminal mutants of SMS1 retain SMS activity, but the capability to synthesize SM is lost upon truncation from TM4 and TM6 in both SMS1 and SMS2 [21].

The SMS activity is cancelled upon deletion of TM6. From these findings appears that the LPP-like motifs at TM6 are critical for the activity of SMS. The SMS enzyme is located mainly in the *cis*-, medial-Golgi [22] and plasma membranes [23, 24].

```
hSMS1     MLSASTMKEVVYWSPKKVADWLLENAMPEYCEPLEHFTGQDLINLTQEDFKKPPLCRVSS 60
hSMS2     ------------------------------------------------------------

hSMS1     DNGQRLLDMIETLKMEHHLEAHKNGHANGHLNIGVDIPTPDGSFSIKIKPNGMPNGYRKE 120
hSMS2     ------------------------------------------------------------

hSMS1     MIKIPMPELERSQYPMEWGKTFLAFLYALSCFVLTTVMISVVHERVPPKEVQPPLPDTFF 180
hSMS2     ------------------------------------------------------------

hSMS1     DHFNRVQWAFSICEINGMILVGLWLIQWLLLKYKSIISRRFFCIVGTLYLYRCITMYVTT 240
hSMS2     -------------------------MTRMFLNNPSIISRRFFCIVGTLYLYRCITMYVTT 35
                                   :  ::*:   *************************

hSMS1     LPVPGMHFNCSPKLFGDWEAQLRRIMKLIAGGGLSITGSHNMCGDYLYSGHTVMLTLTYL 300
hSMS2     LPVPGMHFNCSPKLFGDWEAQLRRIMKLIAGGGLSITGSHNMCGDYLYSGHTVMLTLTYL 95
          ************************************************************

hSMS1     FIKEYSPRRLWWYHWICWLLSVVGIFCILLAHDHYTVDVVVAYYITTRLFWWYHTMANQQ 360
hSMS2     FIKEYSPRRLWWYHWICWLLSVVGIFCILLAHDHYTVDVVVAYYITTRLFWWYHTMANQQ 155
          ************************************************************

hSMS1     VLKEASQMNLLARVWWYRPFQYFEKNVQGIVPRSYHWPFPWPVVHLSRQVKYSRLVNDT 419
hSMS2     VLKEASQMNLLARVWWYRPFQYFEKNVQGIVPRSYHWPFPWPVVHLSRQVKYSRLVNDT 214
          ***********************************************************
```

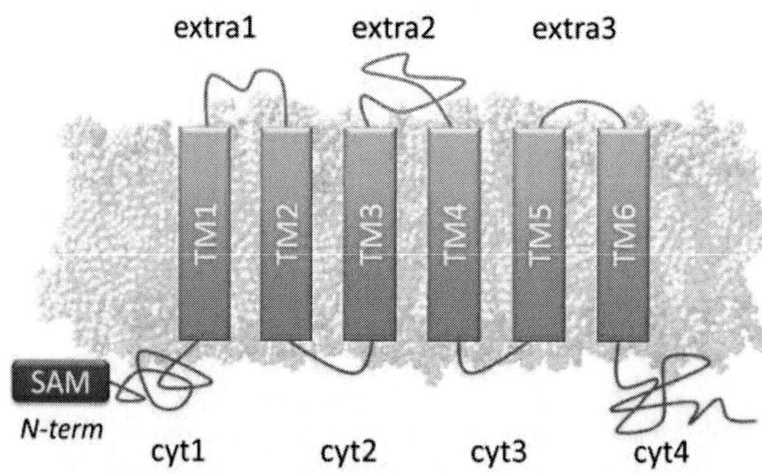

Figure 2. Pair wise alignment of SMS1 and SMS2 (above). Topological representation of SMS1 in cell membrane.

Among the two mammalian SMS gene isoforms, SMS1 is located in the Golgi, whereas SMS2 can be found in both Golgi and plasma membrane [16, 17, 25]. SMS is the last enzyme for SM biosynthesis. Since the enzyme is also able to catalyze the reverse reaction, i.e. the formation of PC from SM and DAG [26], it may regulate the cellular levels of the bioactive lipids Cer and DAG. This is particularly important since Cer and DAG are important regulators of cell proliferation and apoptosis [27]. Since the discovery of the 'SM cycle,' SM has received attention by many research laboratories (Figure 3).

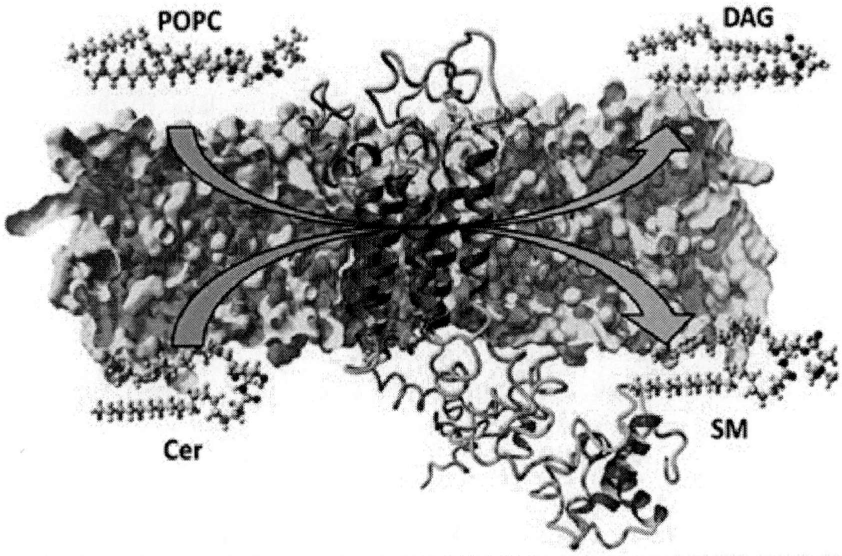

Figure 3. Scheme of the enzymatic reaction regulated by SMS1.

1.3. Membrane Lipid Organization

In the fluid mosaic model, membranes are formed by a fluid bilayer made of proteins and lipids that are free to move [28]. Initially, the mosaic model considered only the random behavior of proteins in the bilayer, but today, the model contemplates additional structural limitations on membrane organization [29, 30]. As matter of fact, most biological membranes are both laterally and transversally asymmetrical. The cross-sectional asymmetry reflects the lipid composition of each leaflet [31].

Sphingomyelin (SM) and phosphatidylcholine (PC) are mostly localized in the outer layer of the plasma membrane. In contrast, the cytoplasmic layer is rich in amine-containing glycerophospholipids such as PE and phosphatidylserine (PS) due to the action of various enzymes [31-35]. In the cytoplasmic layer other phospholipids are present, such as phosphatidylinositol (PI), phosphatidic acid (PA), phosphatidylinositol-4-monophosphate (PIP) and phosphatidylinositol-4,5-biphosphate (PIP2); here they participate in cell signaling [18, 32, 36]. The asymmetric distribution of phospholipid types explains most of the differences between inner and outer layers. As already mentioned, most membranes exhibit also lateral asymmetry. The formation of lateral segregation is the results of chemical-physical differences in lipids and, in some cases, the effect of cytoskeletal structures supporting the membrane, which restrict the movement of both lipids and proteins. The lateral asymmetry is observed both in model systems and in biological membranes [37, 38] and it is mainly due to limited lipid miscibility.

A lipid raft can be defined as a single, closed to planarity, detergent-insoluble glycolipid enriched membrane fraction, with a high content of SM, glycosphingolipids, cholesterol, and proteins. Another kind of membrane asymmetry are caveolae, cell surface invaginations stabilized by structural proteins such as caveolins and lacking in GPI-anchored proteins [39].

Lipids influence the activity of membrane proteins by means of their chemical nature and the physical properties of lipid mixtures in function of their composition [40]. Eukaryotic cells contain a wide variety of membrane lipids, the proportions and types of which define different types of membrane microdomains, regions that regulate: (i) the activity of the proteins; (ii) their localization; and (iii) protein–protein interactions and ensuing signal propagation [41]. Furthermore, during pathological conditions such as cancer [42] or Alzheimer's disease [43] and ageing processes [44, 45] membrane lipid composition can change significantly. Each of these changes causes a membrane remodeling, with consequent modifications of the organization and dynamic properties of lipids. As a consequence, the activities of many membrane-associated proteins and transporters also change dramatically [46-48]. Therefore, it is conceivable that molecules capable of interacting with membrane lipids may induce modifications in membrane composition, protein function or gene expression and reversion of the pathological state.

2. Sphingomyelin in Membrane Organization

2.1. Raft Formation

The high packing density and affinity of sphingomyelin for sterols is responsible for dynamic clustering to form distinct membrane regions, the rafts, which move within the fluid bilayer. These regions were first defined as detergent-resistant transient microdomains formed of cholesterol, sphingomyelin, specific phospholipids with saturated chain, and some embedded proteins [1, 49].

These regions are attracting enormous interest in virtue of their implication in cellular processes like signal transduction, vesicular trafficking and viral infection. The carbonyl group of sphingolipids is involved in hydrogen bonding with the cholesterol hydroxyl group, assuming a conformation with the phospholipid acyl chains fully extended. A sphingolipid molecule has the phosphatidyl-based head group structure but, unlike phospholipid molecules, contains only one fatty acid and a long-chain alcohol as hydrophobic components. The backbone of the sphingolipid is sphingosine, an amino alcohol, rather than glycerol. In 1994, Chong reported that cholesterol molecules tend to distribute regularly and uniformly into superlattices in fluid lipid bilayers [50]. This model proposes that sterols can be organized in a regular lattice within the raft domain. In regular regions, cholesterol molecules are distributed into either hexagonal or rectangular superlattices. The superlattice model is also supported by the findings of molecular dynamics simulations that have revealed an optimal concentration of circa 40% of cholesterol in SM/CHOL domains [51, 52]. Cholesterol is a planar and rigid molecule that clusters preferentially with saturated fatty acid chains, forcing them to form lower energy structures. They can adopt two coexisting fluid phases: liquid-ordered (L_o) and liquid-disordered (L_d). The L_o phase is more rigid and thick than the L_d phase that contains low percentage of cholesterol with unsaturated lipids, producing a final structure less compact and more fluid. This is observed typically in PE enriched domains, where proteins with membrane anchors (e.g., farnesol or geranylgeraniol chains) can easily dock the membranes.

The raft structure strongly depends on the temperature: the distribution and the solubility of proteins in the raft domains could be severely altered by moderate changes in temperature. During raft activation, protein-lipid interactions are coupled to lipid-order–based sorting, generating heterogeneity serving to functionalize, focus, and coordinate the bioactivity of membrane constituents [1].

2.2. Membrane Physical State

In the last 20 years, several experiments showed how cell membranes can sense changes in the surrounding environment. Sudden external changes and/or pathological states can induce membrane lipid reorganization and, consequently, loss of physiological functions [53]. The membrane physical state depends primarily on the lipid composition and is involved in regulating the conformation of embedded proteins and, therefore, their activity. Several essential cellular activities depend on proper membrane state. Thus, membrane lipid composition and membrane lipid dynamics have received significant attention in the past several years [54].

It is well known that when temperature changes, cells produce adaptive responses. For decades, protein denaturation has been supposed to act as a major cause for heat shock protein (Hsp) induction. Recently a new hypothesis suggests the role of the membrane as sensor, in which the remodeling of microdomain and the physical proprieties of the membrane are involved in the heat shock response (HSR) without protein denaturation. Recent studies show that changes in membrane lipid composition and altered heat shock protein levels are often found in some disease such as diabetes, cancer or neurodegenerative diseases. These evidences propose the role of the membrane as a sensor of environmental changes: as a consequence of changes in their phase state and microdomain organization, they transmit signals that activate the expression of genes involved in stress responses [55].

The membrane physical state is often considered as synonymous of fluidity. Fluidity is a concept far too vague to be considered for analyzing the differences among membranes or to quantify the effect of exogenous substances on membrane functionality [55]. In fact, lipid membranes are not isotropic fluids, and the local mobility of lipid tails varies regularly, moving toward the center of the bilayer. Nevertheless, or maybe thanks to the vague definition, the term fluidity is widely used in literature. From an operative point of view, it can be defined in terms of change in fluorescence of a probe embedded in a membrane.

A more sophisticated concept, capable to better capture the complexity of a membrane, is the concept of *lateral stress* or the *lateral pressure profile*. In fact, the interfacial tension between the polar head group region, and the hydrophobic acyl chain region, tends to minimize the surface area, creating a negative pressure at interface. Since a relaxed membrane has zero tension, there must exist a region having a positive pressure (repulsive). The repulsive components are due to electrostatic and steric interactions in the head group and tail regions. The pressure distribution across the membrane is called *lateral pressure profile (LPP).* Cantor [56] presented a theoretical work where he demonstrated that the LPP significantly contributes to the total free energy barrier of a protein. It has been found that the addition of short chain alcohols into membranes drastically alters the LPP and this is the cause of the dissociation of a membrane protein complex (KcSa) [57]. From this evidence, it is clear that membrane can act as sensor of environmental changes. As a consequence of stress, membrane can activate gene expression by changing its physical state and microdomain organization [62]. Fluorescently labeled lipids can evidence the phase separation in lipid rafts, because they can segregate in the liquid disordered domains, due to the steric hindrance of the bulky florescent moieties. To better analyze the interior of the ordered raft domains, Ando et al. [58] introduced in a membrane an SM analog marked with a Raman active diyne moiety, which did not affect the membrane properties, and produced a strong and well visible Raman signal. In contrast to the prediction of the accepted raft model, which assumes a clear distinction between raft and non-raft phases, Raman spectroscopy evidenced a heterogeneous spatial distribution of the diyne-SM probe within raft ordered domains. One explanation proposed was that SM can gradually distribute between SM-reach and SM-poor regions to offset the hydrophobic discrepancy. The authors evidence the ability of hyperspectral Raman imaging to provide information both on the chemical structure and on the spatial distribution of lipids, thus enabling in the future a "quantitative" analysis of lipid membranes. In the past 20 years great attention has been given to membrane lipid composition and dynamics, since membrane functions affect essential cellular activities. Fluidity of membrane is extremely important for its sensitivity to environment: membrane can sense changes in temperature, pH, pressure and it can react to the surroundings by losing physiological functions.

The distribution of L_o and L_d regions in a membrane is a feature of great experimental interest. The position, and in some cases, the thickness and the LPP, of L_o domains is mostly followed by means of fluorescent probes. Nowadays, there is a high number of available probes [59]. The choice of a probe depends on the membrane under investigation and on the fluorescent technique. Fluorescent probes can be classified into 3 classes: (i) probes that bind specifically some lipids, (ii) probes capable to discriminate between L_o and L_d, and (iii) environment sensitive probes. To date, several works have been done to clarify the correlation between the probe structure and its preference to L_o or L_d phases [59-61]. Most probes prefer to bind to L_d phase. This seems to be a general tendency, because highly packed L_o phase are less permeable than L_d phase to exogenous molecules.

2.3. Membrane Fluidity Measurements

Lipid raft can be detected using fluorescent molecular probes. Fluorescence microscopy techniques is a preferred method to monitor membrane composition alterations because it is fast non-invasive, and could be used in situ on live cells. This approach might be used to study the effect of some compounds altering lipid fluidity. Membrane composition alterations outcome in changes of the mobility of the fluorescent probe bound to the membrane. The basic idea is that changes in the lipid composition during changes temperature-dependent or endosome maturation can be monitored by measuring the emitted fluorescence of the probe.

Many probes are available. Environment-sensitive probes are gaining interest because these molecules show a selective partition between L_o and L_d phase. In addition, they change their fluorescence properties giving a different fluorescence signal (color and intensity) in response to environment proprieties, such as hydration, viscosity or polarity [62]. Nile Red is a typical example of environment-sensitive membrane probe [63]. This molecule is modified with long alkyl chain and zwitterionic group to obtain NR12S. The final structure of this analogue constrains the probe on the outer side of the membrane limiting the flip-flop phenomena [64]. NR12S emission wavelength is different between L_o and L_d phase. It is possible to discern between the two phases by monitoring the emission color, thanks to the blue-shift from 605 nm in L_d phase to 570 nm in L_o phase. In GUVs, this membrane probe binds both phases with a preferred binding to the L_d phase providing an excellent contrast in ratiometric images. There are many applications of NR12S, for instance it could be used as a lipid marker to explore the membrane organization during sphingomyelin hydrolysis and cholesterol depletion [64, 65]. Also it has been useful to show the decrease of lipid order in plasma membrane during apoptosis [66]. These studies suggest that apoptosis is linked to a decreased L_o phase in cell membrane. This is probably due to a higher flip-flop rate of SM from the outside leaflet [67] and unsaturated lipids from the inner leaflet.

3. Modulation of the Membrane Physical State

The physical state of membranes is a concept that includes fluidity as well as lateral stress and depends by the membrane composition. The physical state can be altered by external factors such as temperature and pressure, or by small hydrophobic molecules acting on membranes. The sphingomyelin/cholesterol ratio in raft models is critical to shape their interaction, and a threshold concentration of cholesterol (around 40%) may differentiate two different interaction mechanisms [50]. Many evidences gathered during the last two decades show that the protein activity is influenced and, in some cases, modulated by the physical state of membranes. The interest of biologists has focused on the direct role of lipids in influencing the action of membrane proteins [68, 69], the assembly of protein-lipid [70, 71] and the interaction between proteins [70, 71]. After years of intense research, still the exact role played by lipid domains under stress condition is not clear. It was speculated that a possible role of the enormous number of lipid molecular species, over a thousand, is to provide an extra level of regulation for stress signal at the membrane level [8].

Since lipid membranes are kept together by weak molecular interactions, membrane lipids are well adapted to respond to various environmental perturbations. Even limited changes in the composition of some lipid species can alter the overall organization of lipids in terms of fluidity, LPP, lipid-protein interactions and microdomain organizations. Lipid rafts, are particularly sensitive even to slight composition changes. Therefore, physical or chemical stresses can modify the raft organization, and also can change the metabolism of membrane lipids [6, 8, 30, 72, 73].

In the following sections we will review raft domain modulations. In Section 3.1 we will consider the subtle interplay between lipid structure and protein activity. In Section 3.2 and 3.3, we will review the modification of the membrane physical state by exogenous molecules (fluidizers), and by chemical modification of lipids.

3.1. Lipid Mediators of Stress Response

Typical non-raft phospholipids and raft lipid classes like SM, can serve as stress-response modulating factors. It is well known that lipids are reservoirs of arachidonic acid upon action of phospholipase A2. Sphingomyelin has structural and functional roles in the lipid raft, also it is involved in the generation and transduction of Hsp signals. SM can quickly release Cer and sphingosine by the action of sphingomyelin synthase (SMS) under stress conditions [74]. Ceramide, ceramide-1-phosphate, sphingosine and sphingosine-1-phosphate exhibit second messenger functions linked to the cellular Hsp pattern [75].

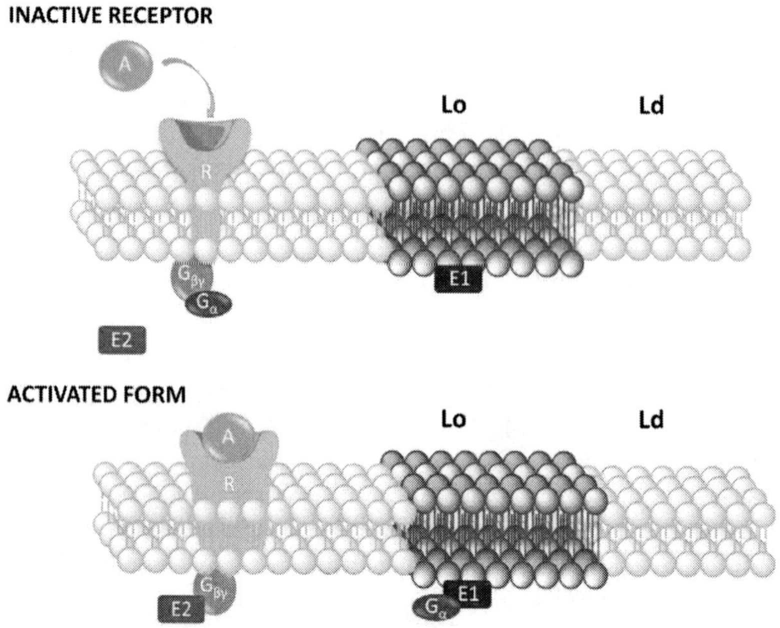

Figure 4. GPCR-signaling. After agonist (A) binding to the GPCRs (R), the G_α subunit moves from the $G_{\beta\gamma}$ dimers to lipid raft (L_o) and here it interacts with specific effector proteins (E1). $G_{\beta\gamma}$ dimers keep the position in L_d phase where can interact with its specific effector (E2). Adapted from [80].

It has been proposed [6, 7, 76] that limited changes of membrane lipids can trigger remarkable modifications in the expression pattern of Hsps, through effects on the membrane physical state, or *via* explicit chemical interactions of lipids with proteins embedded in membrane. The significance of changes in SM levels will be discussed in section 3.3 Lipid modifications. The biophysical properties of the lipid bilayer modulate the activity of membrane proteins. As example one can take the mechanism of action of G proteins. Lipid rafts play a crucial role in the second step of the signaling cascade, since the G_α subunits moves against lipid rafts after the receptor activation (Figure 4). The L_o phase in the membrane definite biophysical properties which allow to trap the activated G_α subunits enhanced the interaction between them and specific effectors. In contrast $G_{\beta\gamma}$ dimers preserve a high affinity for L_d phase [77]. $G_{\beta\gamma}$ dimers are excluded from lipid rafts [78, 79]. This suggests that one probably role of the $G_{\beta\gamma}$ dimer can be linked to the determination the lipid preference of heterotrimeric G proteins. In other words, the $G_{\beta\gamma}$ dimer transports G_α subunits closely to the receptor, supplying them available for an immediate activation.

G protein lipid modifications (fatty acid and/or isoprenyl moieties) are probably involved in this effect, through which the different forms of G proteins are sorted into different membrane domains. The specific localization of membrane proteins is crucial for cell signaling because, in each membrane microdomain, a given molecular entity (e.g., GPCRs, G proteins, etc.) may interact productively with different upstream and downstream signaling proteins. By extension, drugs targeted to membrane lipids can be used to regulate membrane structure and to reverse pathological malfunctions whose etiology is associated with alterations in membrane protein signaling [41, 81].

Increasing the complexity level, there are the effects of membrane remodeling in controlling the heat shock protein response.

Heat-shock proteins (Hsps) play a key role in all live cells. They avoid protein aggregations, assist protein refolding and chaperone other proteins. Different stressors as temperature, infections or hypoxia inflammation induce them to maintain the homeostasis and to supply cellular protection, in stress conditions, cell increases the intracellular levels of Hsps [82]. Hsps identify unfolded and/or damaged protein and assist protein refolding to repair them or active the proteolysis mechanism. Hsps family is divided in different groups accordingly to their molecular weight (Hsp100, Hsp90, Hsp70, Hsp60 and the 'small Hsps,' sHsps) [83].

Membrane lipids adapt their structure and composition in response to several external stresses, such as temperature, pressure, nutrients and other [8]. Some of these changes include modification of the head groups or acyl chains that can modify the total organization of lipids bilayer. In this way the bilayer stability and fluidity can change altering the microdomain organization and ultimately the lipid-protein interactions.

During the last two decades, evidences accumulated showing that the protein activity is influenced and, in some cases, modulated by the physical state of membranes.

3.2. Membrane Fluidizers

In 2005 Vigh and coworkers [84] noticed that the addition of two local anesthetics, benzyl alcohol and heptanol, were capable to induce, in K562 cells, an increase a membrane fluidity similar to that occurring during a thermal shift up to 42°. Associated with this increase in fluidity, they observed an increase in stress gene expression at physiological temperature. Notably, they obtained no evidence of protein unfolding signal. Following this approach, several other molecules have been considered to control gene expression by controlling the membrane fluidity.

Anticancer drugs belonging to the class of hydroxylamine are known to artificially alter the membrane physical state of stressed cells and restore homeostasis. Hydroxylamines are capable to mimic a heat shock stress in raft of sphingomyelin/cholesterol. In particular, BGP-15 and NG-094 are two hydroxylamine derivatives that can induce in membranes an alteration in fluidity similar to those induced by heat [7, 8, 55, 76]. Molecular dynamics analysis revealed that the alteration in atoms mobility in rafts is of several angstroms and that BGP-15 and NG-094 affect differently the surface and the inner part of the membranes. Both molecules can increase the penetration of water in membranes and they induce a rearrangement of SM head groups. Interestingly, these hydroxylamine derivatives have an effect on the fluidity of tens of molecules around them, showing a strong effect on microdomains reorganization.

3.3. Lipid Modifications

Free fatty acid (FFA) can modify the organization of lipid membranes and influence the activity of membrane proteins and signaling processes. Recently, it was demonstrated that non-hydroxylate FA and 2-hydroxylated FA derivatives are capable to reorganize lipid rafts in model membranes [85]. It was also showed that the treatment of lipid vesicles with natural and hydroxylated *cis*-unsaturated FFAs causes a rapid incorporation of the free fatty acids into the lipid bilayers with an increase in membrane fluidity.

The incorporation of FFAs, augment the proportion of L_d structures and reduce the dimension and the rigidity of liquid-ordered domains. Modest alterations in fatty acid structure can effect significant regulatory effects in model lipid membranes [85, 86] and a relevant example is the one offered by 2OHOA, a fatty acid capable of regulating the membrane physical state, the ratio between L_o phase and L_d phase, and the propensity to the non-lamellar phase [87].

Recently, it was demonstrated that 2OHOA activates the sphingomyelin synthases, inducing striking increases, up to 5 fold, in the levels of SM and concomitant decreases in PE (88). An all-atom molecular dynamics analysis of Piotto (51) permitted to clarify the short and long term effect of 2OHOA on L_o and L_d structures.

The L_o and L_d model membranes were made with different proportions of SM, CHOL, POPC and POPE, and the membranes were followed after incorporation of 2OHOA in the bilayer structure (Figure 5). Furthermore, a L_o domain of SM/CHOL immersed into an L_d bilayer of POPC was studied, to estimate the stability of the lipid raft after 2OHOA incorporation in the L_d phase. The results have shown that when the CHOL percentage in SM/CHOL membranes drops below 40%, the water can penetrate the membrane surface because the raft is less compact. The incorporation of 2OHOA into POPC or POPE changes the LPP of the membrane, therefore influencing the binding of membrane proteins, increases the CHOL hydration and increases the bilayer thickness [51].

For L_o/L_d interaction the same authors observed a slight contraction of POPC lipid areas close to the raft [52]. This is accompanied by an increase in thickness of the POPC bilayer. It was observed a reduction of lateral mobility of POPC molecules and a change in the membrane lateral stress profile. The regions in contact with the lipid raft become more "ordered" and the boundary between L_o and L_d regions blurred. Also the stability of the raft is influenced. A SM/CHOL raft immersed in hydroxylated POPC appears unstable and some cholesterol molecules tend to leak out of the raft. On the contrary, a SM/CHOL domain immersed in a POPC bilayer remains stable with no deformation detectable during 50 ns of simulation.

The raft perimeter increases, despite the loss of CHOL molecules and the consequent reduction of the raft volume. In fact, the circular raft undergoes a radical reorganization. Simultaneously, the reduction in CHOL content produces an increase of fluidity in the inner part of the raft. Altogether these data confirm and explain the long term effect of 2OHOA on lipid membranes. It was demonstrated how chemical modification of lipids localized in the L_d region can influence the overall behavior and the physical state of the L_o regions. These findings may offer a reasonable mechanism for understanding the alteration in the membrane receptor activities. The change in morphology and the altered lateral pressure profiles will probably influence the structure and dynamics of membrane proteins.

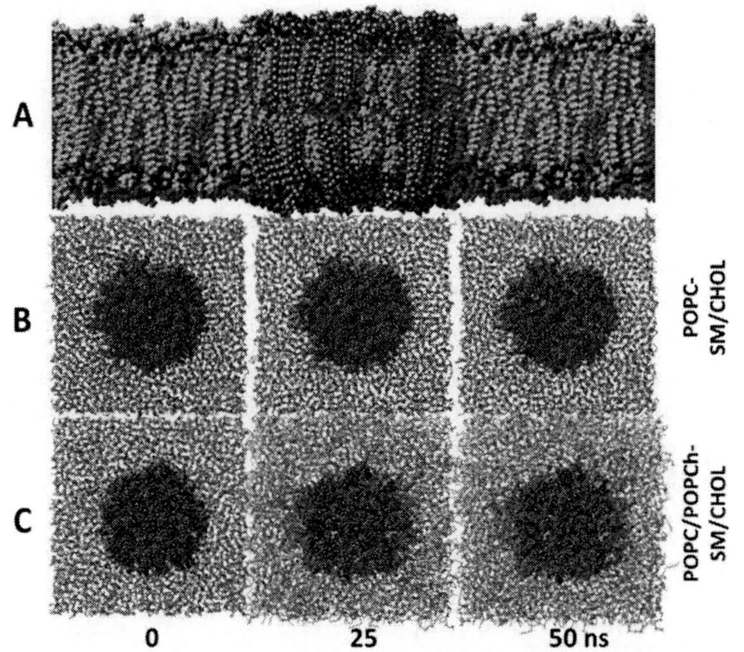

Figure 5. Circular lipid raft of SM:CHOL (60:40) (L_o) immersed in a POPC bilayer (L_d). CHOL carbons are in yellow, SM carbons in blue, and any other atom is colored according CPK standard. Water and counterions are not shown. (A) Lateral view of the system. (B) Top view of time-sequence changes of the SM:CHOL:POPC system after 0, 25 and 50 ns. (C) Time-sequence changes of SM:CHOL lipid raft embedded in a hydroxylated POPC bilayer system after 0, 25 and 50 ns. The raft is not stable, and several CHOL molecules are leaking out into the Ld phase [52].

4. Control of Lipid-Related Enzymes

It has been described that in different human tumor tissues the SM levels are lower compared to normal tissues [89]. The remarkably lower levels of SM in cancer cells suggest that the membrane physical changes of lipid membranes may play a role in the malignant transformation occurring in cancer cells. Experimental data lead to the conclusion that the transformations of glioma and other cancer cells not only require the activation of certain oncoproteins (e.g., Ras) and the inactivation of some tumor suppressor gene, but also, the modification of the membrane structure. In fact the increase in SM diminishes the L_d domains and augments the L_o regions, the net result is an inactivation of the MAPK pathway. Moreover, overexpression of both SMS enzymes promotes TNF-α-mediated apoptosis by a significant increase in SM and DAG levels within Chinese hamster ovary cells. For these reasons, a reasonable therapeutic target can be SMS. In normal cells, high sphingomyelin levels reduce the proliferative capacity of the cell, so that restoration of this situation in cancer cells would be a potential means to treat cancer. In fact, the synthetic fatty acid, 2-hydroxyoleic acid, activates the enzyme sphingomyelin synthase (SMS) and restores membrane sphingomyelin to those levels found in normal cells. While SMS1 and SMS2 deficiency leads to a considerable reduction in SM and DAG levels within the THP-1-derived macrophages and contributes to apoptosis induced by lipopolysaccharide [90]. From these results it is clear that both SMS1 and SMS2 regulate intracellular SM and DAG levels and could contribute to changes in lipid rafts on the plasma membrane. Human SMS is a transmembrane enzyme with cytosolic *N*- and *C*-terminal portions. The terminal Sterile Alpha Motif (SAM) domain does not influence SMS activity. SMS mutants with deletion of SAM domain from SMS1 or addition of SAM to SMS2 showed no appreciable impact on SMS activity (16]. SM synthases is very similar to the lipid phosphate phosphatase (LiPP). So it is realistic that they share a common catalytic mechanism [17]. The LiPP family is involved in cell signaling by regulating the conversion of bioactive lipid phosphate ester to their dephosphorylated counterparts. SMS enzymes show four highly conserved sequence motifs and two of them are close to the phosphatase domains in LiPPs containing the catalytic histidine and aspartate residues involved in enzymatic reaction [16, 91]. This idea was supported by specific activity assay on five SMS1mutants in which each amino acid of the proposed catalytic site was replaced with alanine and showed no detectable activity [16]. Actually, the main accredited hypothesis is that the HHD triad, in the transmembrane portion, is responsible for SMS activity.

Conclusion

Membranes are the "meeting point" of many proteins and lipids; in membrane structures, the majority of cellular activities occur. Sphingolipids are usually found associated with sterols to form dynamic lateral membrane rafts. The control of the raft composition and physical state is crucial in regulating stress or inflammatory responses, protein mobility and clustering, and protein-protein communications. Steadily increasing experimental results indicate that raft domains of SM/CHOL play multiple and interconnected crucial roles in cell life, such as stress sensors, signal amplification, and control of protein activity. At the same time, several attempts have been made to control the physical state of these regions. Several experiments exploit exogenous molecules to increase the fluidity of membranes and, in this way, reproduce the effect of a mild heat shock. Other works concentrated to the cell treatment with modified fatty acids to achieve a long term modification of the bilayers. Finally, the levels of sphingomyelin can be controlled targeting the sphingomyelin synthase. It is now evident that the control of the domain composition and/or its physical state is extremely promising for developing more effective molecular therapies of lipid related diseases.

References

[1] Lingwood D, Simons K. Lipid Rafts As a Membrane-Organizing Principle. *Science*. 2010;327(5961):46-50.

[2] Lippincott-Schwartz J, Phair RD. Lipids and Cholesterol as Regulators of Traffic in the Endomembrane System. *Annual Review of Biophysics*. 2010;39:559-78.

[3] Harder T. Formation of functional cell membrane domains: the interplay of lipid- and protein-mediated interactions. *Philosophical Transactions of the Royal Society B: Biological Sciences*. 2003;358(1433):863-8.

[4] Maula T, Westerlund B, Slotte JP. Differential ability of cholesterol-enriched and gel phase domains to resist benzyl alcohol-induced fluidization in multilamellar lipid

vesicles. *Biochimica et Biophysica Acta (BBA) - Biomembranes*. 2009;1788(11): 2454-61.

[5] Nagy E, Balogi Z, Gombos I, Åkerfelt M, Björkbom A, Balogh G, et al. Hyperfluidization-coupled membrane microdomain reorganization is linked to activation of the heat shock response in a murine melanoma cell line. *Proceedings of the National Academy of Sciences of the United States of America*. 2007;104(19):7945-50.

[6] Vigh L, Escribá PV, Sonnleitner A, Sonnleitner M, Piotto S, Maresca B, et al. The significance of lipid composition for membrane activity: new concepts and ways of assessing function. *Progress in lipid research*. 2005;44(5):303-44.

[7] Gombos I, Crul T, Piotto S, Gungor B, Torok Z, Balogh G, et al. Membrane-lipid therapy in operation: the HSP co-inducer BGP-15 activates stress signal transduction pathways by remodeling plasma membrane rafts. *PloS one*. 2011;6(12):e28818.

[8] Vigh L, Torok Z, Balogh G, Glatz A, Piotto S, Horvath I. Membrane-regulated stress response: a theoretical and practical approach. *Advances in experimental medicine and biology*. 2007;594:114-31.

[9] Hannun YA, Obeid LM. Principles of bioactive lipid signalling: lessons from sphingolipids. *Nature reviews Molecular cell biology*. 2008;9(2):139-50.

[10] Young MM, Kester M, Wang H-G. Sphingolipids: regulators of crosstalk between apoptosis and autophagy. *Journal of lipid research*. 2013;54(1):5-19.

[11] Kitatani K, Idkowiak-Baldys J, Hannun YA. The sphingolipid salvage pathway in ceramide metabolism and signaling. *Cellular signalling*. 2008;20(6):1010-8.

[12] Thomas DM, Yusuf AH, Lina MO. Ceramide synthases at the centre of sphingolipid metabolism and biology. *Biochemical Journal*. 2012;441(3):789-802.

[13] Marchesini N, Hannun YA. Acid and neutral sphingomyelinases: roles and mechanisms of regulation. *Biochemistry and cell biology*. 2004;82(1):27-44.

[14] Luberto C, Hannun YA. Sphingomyelin synthase, a potential regulator of intracellular levels of ceramide and diacylglycerol during SV40 transformation. Does sphingomyelin synthase account for the putative phosphatidylcholine-specific phospholipase C? *The Journal of biological chemistry*. 1998;273(23):14550-9.

[15] Yamaoka S, Miyaji M, Kitano T, Umehara H, Okazaki T. Expression cloning of a human cDNA restoring sphingomyelin synthesis and cell growth in sphingomyelin synthase-defective lymphoid cells. *Journal of Biological Chemistry*. 2004; 279(18):18688-93.

[16] Yeang C, Varshney S, Wang R, Zhang Y, Ye D, Jiang X-C. The domain responsible for sphingomyelin synthase (SMS) activity. *Biochimica et Biophysica Acta (BBA)-Molecular and Cell Biology of Lipids*. 2008;1781(10):610-7.

[17] Huitema K, van den Dikkenberg J, Brouwers JF, Holthuis J. Identification of a family of animal sphingomyelin synthases. *The EMBO journal*. 2004;23(1):33-44.

[18] Bork P, Koonin EV. Predicting functions from protein sequences—where are the bottlenecks? *Nature genetics*. 1998;18(4):313-8.

[19] Barrera FN, Poveda JA, González-Ros JM, Neira JL. Binding of the C-terminal sterile α motif (SAM) domain of human p73 to lipid membranes. *Journal of Biological Chemistry*. 2003;278(47):46878-85.

[20] Qiao F, Bowie JU. The many faces of SAM. *Science Signaling*. 2005;2005(286):re7-re.

[21] Taniguchi M, Okazaki T. The role of sphingomyelin and sphingomyelin synthases in cell death, proliferation and migration—from cell and animal models to human disorders. *Biochimica et Biophysica Acta (BBA)-Molecular and Cell Biology of Lipids.* 2014;1841(5):692-703.

[22] Jeckel D, Karrenbauer A, Birk R, Schmidt RR, Wieland F. Sphingomyelin is synthesized in the cis Golgi. *FEBS letters.* 1990;261(1):155-7.

[23] Malgat M, Maurice A, Baraud J. Sphingomyelin and ceramide-phosphoethanolamine synthesis by microsomes and plasma membranes from rat liver and brain. Journal of lipid research. 1986;27(3):251-60.

[24] Moreau P, Cassagne C. Phospholipid trafficking and membrane biogenesis. *Biochimica et Biophysica Acta (BBA)-Reviews on Biomembranes.* 1994;1197(3):257-90.

[25] Yeang C, Ding T, Chirico WJ, Jiang X-C. Subcellular targeting domains of sphingomyelin synthase 1 and 2. *Nutrition & metabolism.* 2011;8(1):1-9.

[26] Marggraf W-D, Kanfer JN. The phosphorylcholine acceptor in the phosphatidylcholine: ceramide cholinephosphotransferase reaction is the enzyme a transferase or a hydrolase? *Biochimica et Biophysica Acta (BBA)-Lipids and Lipid Metabolism.* 1984;793(3):346-53.

[27] Pettus BJ, Chalfant CE, Hannun YA. Ceramide in apoptosis: an overview and current perspectives. *Biochimica et Biophysica Acta (BBA)-Molecular and Cell Biology of Lipids.* 2002;1585(2):114-25.

[28] Singer S, Nicolson GL. The fluid mosaic model of the structure of cell membranes. *Day and Good Membranes and viruses in immunopathology.* 1972:7-47.

[29] Escribá PV, Ozaita A, Ribas C, Miralles A, Fodor E, Farkas T, et al. Role of lipid polymorphism in G protein-membrane interactions: nonlamellar-prone phospholipids and peripheral protein binding to membranes. *Proceedings of the National Academy of Sciences.* 1997;94(21):11375-80.

[30] Vereb G, Szöllősi J, Matko J, Nagy P, Farkas T, Vigh L, et al. Dynamic, yet structured: the cell membrane three decades after the Singer–Nicolson model. *Proceedings of the National Academy of Sciences.* 2003;100(14):8053-8.

[31] Verkleij AJ, Zwaal RF, Roelofsen B, Comfurius P, Kastelijn D, van Deenen LL. The asymmetric distribution of phospholipids in the human red cell membrane. A combined study using phospholipases and freeze-etch electron microscopy. *Biochim Biophys Acta.* 1973;323(2):178-93.

[32] Bretscher MS. Phosphatidyl-ethanolamine: differential labelling in intact cells and cell ghosts of human erythrocytes by a membrane-impermeable reagent. *J Mol Biol.* 1972;71(3):523-28.

[33] Op den Kamp JAF. Lipid asymmetry in membranes. *Annual Review of Biochemistry.* 1979;48:47-71.

[34] Rothman JE, Lenard J. Membrane Asymmetry. The nature of membrane asymmetry provides clues to the puzzle of how membranes are assembled. *Science.* 1977;195:743-53.

[35] van Meer G, Voelker DR, Feigenson GW. Membrane lipids: where they are and how they behave. *Nat Rev Mol Cell Biol.* 2008;9(2):112-24.

[36] Gascard P, Tran D, Sauvage M, Sulpice J-C, Fukami K, Takenawa T, et al. Asymmetric distribution of phosphoinositides and phosphatidic acid in the human erythrocyte membrane. *Biochimica et Biophysica Acta (BBA)-Biomembranes.* 1991;1069(1):27-36.

[37] Mabrey S, Mateo PL, Sturtevant JM. High-sensitivity scanning calorimetric study of mixtures of cholesterol with dimyristoyl-and dipalmitoylphosphatidylcholines. *Biochemistry.* 1978;17(12):2464-8.

[38] Marguet D, Lenne PF, Rigneault H, He HT. Dynamics in the plasma membrane: how to combine fluidity and order. *The EMBO journal.* 2006;25(15):3446-57.

[39] Fielding CJ, Fielding PE. Relationship between cholesterol trafficking and signaling in rafts and caveolae. *Biochimica et Biophysica Acta (BBA)-Biomembranes.* 2003;1610(2):219-28.

[40] Dowhan W. Molecular basis for membrane phospholipid diversity: why are there so many lipids? *Annu Rev Biochem.* 1997;66:199-232.

[41] Escribà PV, González-Ros JM, Goñi FM, Kinnunen PK, Vigh L, Sánchez-Magraner L, et al. Membranes: a meeting point for lipids, proteins and therapies. *Journal of cellular and molecular medicine.* 2008;12(3):829-75.

[42] Fermor BF, Masters JR, Wood CB, Miller J, Apostolov K, Habib NA. Fatty acid composition of normal and malignant cells and cytotoxicity of stearic, oleic and sterculic acids in vitro. *Eur J Cancer.* 1992;28A(6-7):1143-7.

[43] Nakada T, Kwee IL, Ellis WG. Membrane fatty acid composition shows Δ6-desaturase abnormalities in Alzheimer's disease. *NeuroReport.* 1990;1:153-5.

[44] Alvarez E, Ruiz-Gutiérrez V, Santa María C, Machado A. Age-dependent modification of lipid composition and lipid structural order parameter of rat peritoneal macrophage membranes. *Mech Ageing Dev.* 1993;71:1-12.

[45] Parasassi T, Di Stefano M, Ravagnan G, Sapora O, Gratton E. Membrane aging during cell growth ascertained by Laurdan generalized polarization. *Exp Cell Res.* 1992;202:432-9

[46] Avery SV, Lloyd D, Harwood JL. Temperature-dependent changes in plasma-membrane lipid order and the phagocytotic activity of the amoeba Acanthamoeba castellanii are closely correlated. *Biochem J.* 1995;312:811-6.

[47] Hu Q, Moerman EJ, Goldstein S. Altered Expression and Regulation of the alpha5beta1 Integrin-Fibronectin Receptor Lead to Reduced Amounts of Functional alpha5beta1 Heterodimer on the Plasma *Membrane of Senescent Human Diploid Fibroblasts Exp Cell Res.* 1996;224:251-63.

[48] Shmeeda HR, Golden EB, Barenholz Y. Membrane lipids and aging. In: Shinitzky Me, editor. Handbook of Biomembranes Mammalian Membrane, Structure and Function. 2. Weinheim: Balaban VCH, New York; 1994. p. 1-82.

[49] Simons K, Ikonen E. Functional rafts in cell membranes. *Nature.* 1997;387(6633):569-72.

[50] Somerharju P, Virtanen JA, Cheng KH, Hermansson M. The superlattice model of lateral organization of membranes and its implications on membrane lipid homeostasis. *Biochimica et Biophysica Acta (BBA)-Biomembranes.* 2009;1788(1):12-23.

[51] Piotto S, Concilio S, Bianchino E, Iannelli P, Lopez DJ, Teres S, et al. Differential effect of 2-hydroxyoleic acid enantiomers on protein (sphingomyelin synthase) and lipid (membrane) targets. *Biochim Biophys Acta.* 2014;1838(6):1628-37.

[52] Piotto S, Trapani A, Bianchino E, Ibarguren M, Lopez DJ, Busquets X, et al. The effect of hydroxylated fatty acid-containing phospholipids in the remodeling of lipid membranes. *Biochim Biophys Acta.* 2014;1838(6):1509-17.

[53] Cossins AR. Temperature Adaptations of Biological Membranes: London and Chapel Hill: Portland Press; 1994.
[54] Quinn PJ, Joo F, Vigh L. The Role of Unsaturated Lipids in Membrane-Structure and Stability. *Prog Biophys Mol Bio*. 1989;53(2):71-103.
[55] Crul T, Toth N, Piotto S, Literati-Nagy P, Tory K, Haldimann P, et al. Hydroximic acid derivatives: pleiotropic Hsp co-inducers restoring homeostasis and robustness. *Current pharmaceutical design*. 2013;19(3):309-46.
[56] Cantor RS. The influence of membrane lateral pressures on simple geometric models of protein conformational equilibria. *Chemistry and physics of lipids*. 1999;101(1):45-56.
[57] van den Brink-van der Laan E, Chupin V, Killian JA, de Kruijff B. Small alcohols destabilize the KcsA tetramer via their effect on the membrane lateral pressure. *Biochemistry*. 2004;43(20):5937-42.
[58] Ando J, Kinoshita M, Cui J, Yamakoshi H, Dodo K, Fujita K, et al. Sphingomyelin distribution in lipid rafts of artificial monolayer membranes visualized by Raman microscopy. *Proc Natl Acad Sci U S A*. 2015;112(15):4558-63.
[59] Klymchenko AS, Kreder R. Fluorescent probes for lipid rafts: from model membranes to living cells. *Chemistry & biology*. 2014;21(1):97-113.
[60] Sengupta P, Hammond A, Holowka D, Baird B. Structural determinants for partitioning of lipids and proteins between coexisting fluid phases in giant plasma membrane vesicles. *Biochimica et Biophysica Acta* (BBA)-Biomembranes. 2008;1778(1):20-32.
[61] Sezgin E, Levental I, Grzybek M, Schwarzmann G, Mueller V, Honigmann A, et al. Partitioning, diffusion, and ligand binding of raft lipid analogs in model and cellular plasma membranes. *Biochimica et Biophysica Acta (BBA)-Biomembranes*. 2012;1818(7):1777-84.
[62] Demchenko AP, Mely Y, Duportail G, Klymchenko AS. Monitoring biophysical properties of lipid membranes by environment-sensitive fluorescent probes. *Biophys J*. 2009;96(9):3461-70.
[63] Greenspan P, Mayer EP, Fowler SD. Nile red: a selective fluorescent stain for intracellular lipid droplets. *The Journal of cell biology*. 1985;100(3):965-73.
[64] Kucherak OA, Oncul S, Darwich Z, Yushchenko DA, Arntz Y, Didier P, et al. Switchable Nile Red-Based Probe for Cholesterol and Lipid Order at the Outer Leaflet of Biomembranes. *J Am Chem Soc*. 2010;132(13):4907-16.
[65] Darwich Z, Klymchenko AS, Kucherak OA, Richert L, Mely Y. Detection of apoptosis through the lipid order of the outer plasma membrane leaflet. *Bba-Biomembranes*. 2012;1818(12):3048-54.
[66] Gibbons E, Pickett KR, Streeter MC, Warcup AO, Nelson J, Judd AM, et al. Molecular details of membrane fluidity changes during apoptosis and relationship to phospholipase A(2) activity. *Bba-Biomembranes*. 2013;1828(2):887-95.
[67] Tepper AD, Ruurs P, Wiedmer T, Sims PJ, Borst J, van Blitterswijk WJ. Sphingomyelin hydrolysis to ceramide during the execution phase of apoptosis results from phospholipid scrambling and alters cell-surface morphology. *The Journal of cell biology*. 2000;150(1):155-64.
[68] de Kruijff B. Lipids beyond the bilayer. *Nature*. 1997;386(6621):129-30.
[69] Tate MW, Eikenberry EF, Turner DC, Shyamsunder E, Gruner SM. Nonbilayer phases of membrane lipids. *Chemistry and physics of lipids*. 1991;57:147-64.

[70] Gil T, Ipsen JH, Mouritsen OG, Sabra MC, Sperotto MM, Zuckermann MJ. Biochem Biophys Acta. *Theoretical analysis of protein organization in lipid membranes.* 1998;1376:245-66.
[71] Mouritsen OG. Self-assembly and organization of lipid-protein membranes. *Curr Opin Coll Int Sci.* 1998;3(78-87).
[72] Sõti C, Nagy E, Giricz Z, Vígh L, Csermely P, Ferdinandy P. Heat shock proteins as emerging therapeutic targets. *British journal of pharmacology.* 2005;146(6):769-80.
[73] Vigh L, Horváth I, Maresca B, Harwood JL. Can the stress protein response be controlled by 'membrane-lipid therapy'? *Trends in biochemical sciences.* 2007;32(8):357-63.
[74] Jenkins G. The emerging role for sphingolipids in the eukaryotic heat shock response. *Cellular and Molecular Life Sciences CMLS.* 2003;60(4):701-10.
[75] Mathieson FA, Nixon GF. Sphingolipids differentially regulate mitogen-activated protein kinases and intracellular Ca2+ in vascular smooth muscle: effects on CREB activation. *British journal of pharmacology.* 2006;147(4):351-9.
[76] Torok Z, Crul T, Maresca B, Schutz GJ, Viana F, Dindia L, et al. Plasma membranes as heat stress sensors: From lipid-controlled molecular switches to therapeutic applications. *Biochim Biophys Acta.* 2014;1838(6):1594-618.
[77] Vögler O, Casas J, Capò D, Nagy T, Borchert G, Martorell Gea. The Gbetagamma dimer drives the interaction of heterotrimeric Gi proteins with nonlamellar membrane structures. *The Journal of biological chemistry.* 2004;279:36540-5.
[78] Melkonian KA, Ostermeyer AG, Chen JZ, Roth MG, Brown DA. Role of lipid modifications intargeting proteins to detergent-resistantmembrane rafts. Many raft proteins are acylated, while feware prenylated. *The Journal of biological chemistry.* 1999;274:3910-7.
[79] Moffett S, Brown DA, Linder ME. Lipid-dependent targeting of G proteins into rafts. *The Journal of biological chemistry.* 2000;275:2191-8.
[80] Vögler O, Barcelò JM, Ribas C, Escribà PV. Membrane interactions of G proteins and other related proteins. *Biochim Biophys Acta.* 2008;1778:1640-52.
[81] Escribà PV. Membrane-lipid therapy: a new approach in molecular medicine. *Trends in Molecular Medicine.* 2006;12(1):34-43.
[82] Kültz D. Molecular and evolutionary basis of the cellular stress response. *Annu Rev Physiol.* 2005;67:225-57.
[83] Kim HP, Morse D, Choi AM. Heat-shock proteins: new keys to the development of cytoprotective therapies. 2006.
[84] Balogh G, Horváth I, Nagy E, Hoyk Z, Benkõ S, Bensaude O, et al. The hyperfluidization of mammalian cell membranes acts as a signal to initiate the heat shock protein response. *FEBS Journal.* 2005;272(23):6077-86.
[85] Ibarguren M, Lopez DJ, Encinar JA, Gonzalez-Ros JM, Busquets X, Escriba PV. Partitioning of liquid-ordered/liquid-disordered membrane microdomains induced by the fluidifying effect of 2-hydroxylated fatty acid derivatives. *Biochim Biophys Acta.* 2013;1828(11):2553-63.
[86] Funari SS, Barcelo F, Escriba PV. Effects of oleic acid and its congeners, elaidic and stearic acids, on the structural properties of phosphatidylethanolamine membranes. *J Lipid Res.* 2003;44(3):567-75.

[87] Barcelo F, Prades J, Funari SS, Frau J, Alemany R, Escriba PV. The hypotensive drug 2-hydroxyoleic acid modifies the structural properties of model membranes. *Molecular membrane biology*. 2004;21(4):261-8.

[88] Terés S, Lladó V, Higuera M, Barceló-Coblijn G, Martin ML, Noguera-Salvà MA, et al. 2-Hydroxyoleate, a nontoxic membrane binding anticancer drug, induces glioma cell differentiation and autophagy. *Proceedings of the National Academy of Sciences of the United States of America*. 2012;109(22):8489-94.

[89] Hendrich AB, Michalak K. Lipids as a target for drugs modulating multidrug resistance of cancer cells. *Curr Drug Targets*. 2003;4(1):23-30.

[90] Ding T, Li Z, Hailemariam T, Mukherjee S, Maxfield FR, Wu MP, et al. SMS overexpression and knockdown: impact on cellular sphingomyelin and diacylglycerol metabolism, and cell apoptosis. *J Lipid Res*. 2008;49(2):376-85.

[91] Zhang QX, Pilquil CS, Dewald J, Berthiaume LG, Brindley DN. Identification of structurally important domains of lipid phosphate phosphatase-1: implications for its sites of action. *Biochem J*. 2000;345 Pt 2:181-4.

In: Sphingolipids
Editor: Angel Catalá

ISBN: 978-1-63483-019-5

Chapter 2

TRICYCLIC ANTIDEPRESSANTS ARE MODULATORS OF SPHINGOLIPID TURNOVER AND CELL SENSITIVITY TO INSULIN ACTION

Nataliya A. Babenko*
Department of Physiology of Ontogenesis, Institute of Biology,
Kharkov Karazin National University, Kharkov, Ukraine

ABSTRACT

Because of their effect on the reuptake of norepinephrine or serotonin tricyclic antidepressants are used to treat the depressive disorders. Such antidepressants as amitriptyline, desipramine, imipramine, nortriptyline, doxepin, clomipramine, and protriptyline are the more commonly used. Findings from cellular and animal studies demonstrated that antidepressant drugs may regulate ceramide levels and decrease acid sphingomyelinase (aSMase) activity. Such tricyclic antidepressants as desipramine and imipramine cause via enzyme proteolytic degradation a detachment of the aSMase from the inner membrane and subsequent inactivation. These drugs do not inhibit aSMase directly, but may act as functional inhibitors of SMases and thus produce the anti-apoptotic, proliferative and anti-inflammatory effects. Moreover, the aSMase inhibitor, amitriptyline, attenuated obesity and obesity-associated renal injury, induced by the high-fat-diet treatment. Inhibition of aSMase by desipramine significantly reduced triacylglycerol synthesis and content and increased the (dihidro)ceramide synthesis in the HepG2 cells, cultured in the presence of exogenous palmitic acid. These results provide direct evidence that aSMase supplies the accumulated palmitic acid mainly to sphingolipid synthesis. In contrast, the overexpression of aSMase in the diabetic db/db mice is associated with glucose and lipid metabolism improvement. aSMase inhibition with imipramine decreases glucose uptake by liver cells and glycogen and lipid deposition in hepatocytes. However, inhibition of aSMase by imipramine partly improves the hepatocyte sensitivity to insulin action in the aged rats. Treatment of "old" insulin-resistant cells with imipramine reduces the elevated aSMase activity and ceramide synthesis and increases insulin-induced glucose uptake and glycogen synthesis. Although the functional inhibitors of aSMase are widely used in cellular and animal studies to

* E-mail: babenko@univer.kharkov.ua

improve glucose and lipid homeostasis, the therapeutic consequences of functional inhibition of aSMase are yet poorly understood. However, most of the available tricyclic antidepressants are licensed for medical use in humans, as being minimally toxic and therefore may be used for treatment of disease states associated with increased activity of aSMase. Recently functional inhibitors of aSMase have been shown to reduce the severity of different animal models of human diseases, including cystic fibrosis, Wilson disease, acute lung injury and hepatic ischaemia-reperfusion injury. This chapter will be focused on the role of tricyclic antidepressants in the regulation of sphingolipids turnover in the cells at different pathological conditions associated with enhanced activity of aSMase. Much attention will be given to the modulation of cell sensitivity to insulin action by functional inhibitors of aSMase activity.

Keywords*:* tricyclic antidepressants, acid sphingomyelinase, ceramide, liver, muscle, brain, insulin resistance

1. INTRODUCTION

Sphingolipids are components of biological membranes and a class of biologically active molecules involved in regulating such physiological processes as cell growth, apoptosis, autophagy, angiogenesis, inflammation, and neurodegeneration. Sphingomyelin is the most abundant sphingolipid and its hydrolysis under the action of sphingomyelinases is the important pathway of ceramide generation in cells. The sphingomyelin metabolism is considered to be a novel signal transduction pathway that generates multiple lipid messengers: ceramide, sphingosine, and sphingosine-1-phosphate [1, 2]. The oxidative and metabolic stresses and other factors which altering the metabolism of sphingolipids, induce ceramide accumulation and can result in age-related diseases, as cancers, diabetes, atherosclerosis, and neuro-degenerative disorders [2]. Ceramide leads to membrane reorganization and downstream signaling that results in cell stress or apoptosis. Ceramide is further metabolized to sphingosine and sphingosine-1-phosphate by acid ceramidase and sphingosine kinases, respectively. While ceramide and sphingosine are pro-apoptotic lipids, sphingosine-1-phosphate has been shown to be involved in regulation of cellular differentiation, proliferation and cell migration [3-7]. Thus, balance between ceramide/sphingosine and sphingosine-1-phosphate is extremely important to maintain the balance between growth and cell death [8-10].

Homeostasis of ceramide and sphingosine-1-phosphate in cells is regulated with acid sphingomyelinase. During acid sphingomyelinase activation the balance between these lipids could be inclined toward ceramide or sphingosine-1-phosphate, resulting in death/inflammation or protein kinase Akt activation and cell survival, respectively. For example, using the knockout animals and RNA interference it has been determined that acid sphingomyelinase via sphingosine-1-phosphate formation [11] mediates the insulin-like effects, such as glucose uptake, glycogen deposition, and lipid accumulation in liver cells, while ceramide, accumulated in insulin responsive tissues, alters insulin signaling [12-14], and thereby contributes to insulin resistance [15, 16] and to type 2 diabetes [16]. It should be mentioned that pharmacologic acid sphingomyelinase inhibition imitates effects of enzyme deficiency in hepatocytes isolated from acid sphingomyelinase$^{-/-}$ mice [11]. However, cell lines and animals lacking acid sphingomyelinase develop a sphingomyelin storage phenotype with numerous disturbances of cell metabolism and function. Thus, the pharmacologic inhibitors of acid sphingomyelinase may be important tools for identifying the new role of sphingolipids in cell signaling.

In this chapter, the tricyclic antidepressants are considered as acid sphingomyelinase inhibitors. The role of tricyclic antidepressants in the regulation of sphingolipids turnover in the cells at different pathological conditions is discussed. Much attention is given to the modulation of cell sensitivity to insulin action by functional inhibitors of acid sphingomyelinase activity at old age.

2. Tricyclic Antidepressants As Acid Sphingomyelinase Inhibitors

As tricyclic antidepressants influence on the reuptake of norepinephrine or serotonin, therefor they are commonly used to treat the depressive disorders. Amitriptyline, desipramine, imipramine, nortriptyline, doxepin, clomipramine, and protriptyline are the frequently used antidepressants. The tricyclic antidepressants are well absorbed orally and undergo metabolism in the liver. Findings from cellular and animal studies demonstrated that antidepressant drugs can regulate ceramide levels and decrease acid sphingomyelinase activity [17-19]. It has been demonstrated that acid sphingomyelinase-dependent ceramide accumulation in the hippocampus is of critical importance for the development of depression-like phenotype [20]. Using mice transgenic for acid sphingomyelinase or acid sphingomyelinase deficient mice or treatment by amitriptyline or fluoxetine, the key role of ceramide as a negative regulator of major depression has been suggested [21, 22]. The reduction of acid sphingomyelinase expression and activity in acid sphingomyelinase transgenic mice by amitriptyline or fluoxetine decreased ceramide concentration in the hippocampus and normalized behavior, neurogenesis, neuronal maturation and neuronal survival [20].

Acid sphingomyelinase was the first cloned sphingomyelin phosphodiesterase. The gene that encodes acid sphingomyelinase gives rise to two unique enzymes: the lysosomal sphingomyelinase [23] and a secretory sphingomyelinase [24]. The acid sphingomyelinase (optimum pH 5.0, EC 3.1.4.12) is an intralysosomal glycoprotein, catalyzing the degradation of sphingomyelin. Although there exist other alternative pathways of ceramide production, activation of acid sphingomyelinase itself is of crucial importance for some cellular responses. Translocation of lysosomal acid sphingomyelinase onto the outer leaflet of the cell membrane plays an important role in cellular stress response [25, 26]. Due to increased acid sphingomyelinase activity ceramide accumulation is a common feature of cells treated with CD95 ligands and such cytokines as tumor necrosis factor-alpha (TNF-alpha), interleukin-1beta and interferon-gamma, reactive oxygen and nitrogen species, ionizing radiation, heat shock and other agents of stress, injury or infections [26-31]. Due to decreased ceramide generation acid sphingomyelinase knockout mice are protected against a variety of stress stimuli, including Fas ligand [32], lipopolysaccharide [33], ionizing radiation [34, 35], photocytotoxicity [36], ischemia/reperfusion injury [37], cisplatin [38], and TNF-alpha [39, 40].

Acid sphingomyelinase inhibitors are widely used to protect cells from the stress-induced and receptor-mediated apoptosis and cell malfunction and to promote the cell proliferation [17-19]. Tricyclic antidepressants, desipramine and imipramine, cause a detachment of the acid sphingomyelinase from the inner membrane and subsequent enzyme inactivation via proteolytic degradation. These drugs are not responsible directly for inhibition of acid sphingomyelinase, but lead to its functional inhibition and are called FIASMAs (functional inhibitors of acid sphingomyelinase) [17]. Amitriptyline attenuated obesity and obesity-associated renal injury induced with the high-fat-diet treatment [41]. Inhibition of acid sphingomyelinase by desipramine significantly reduced triacylglycerol synthesis and content, stimulated with palmitic acid, and increased the (dihidro)ceramide synthesis in the HepG2 cells, while myriocin or fumonisin B 1 (ceramide synthesis inhibitors) attenuated the desipramine effect on ceramide synthesis [42]. Similar findings were obtained for acid sphingomyelinase-deficient mice fed with saturated fat enriched-diet. These results provide direct evidence that acid sphingomyelinase supplies palmitic acid mainly to sphingolipid synthesis. To the contrary, the overexpression of acid sphingomyelinase in the diabetic db/db mice is associated with glucose and lipid metabolism improvement [11]. Acid sphingomyelinase inhibition with imipramine decreased glucose uptake by liver cells and glycogen and lipid deposition in hepatocytes. These results clearly demonstrated the important role of acid sphingomyelinase in glucose and lipids metabolism regulation and proved that tricyclic antidepressants can be useful tools in determination of acid sphingomyelinase-dependent pathway in cellular metabolism regulation.

However, it is worth noting that the desipramine could inhibit the acid ceramidase in addition to acid sphingomyelinase in the human prostate cancer cell line DU145, bladder cancer cell line 5637, and Hela cervical cancer cells [43], though this drug has no effect on a majority of other lisosomal enzymes. The recent findings that desipramine induces proteolysis of mature acid ceramidase [43, 44] are of particular importance as suggesting the coordination of lysosomal sphingomyelinase and ceramide salvage pathway regulation [45]. Lysosomal sphingomyelinase may act upstream in the ceramide salvage pathway initiating the ceramide production, followed by processing of acid ceramidase and consequent sphingosine accumulation. Due to its water solubility, sphingosine can escape the lysosome and be reacylated in different cellular compartments to ceramide by ceramide synthases [46]. Thus, it became evident that desipramine could disrupt the both acid sphingomyelinase- and acid ceramidase-dependent pathways and limit production of ceramide by ceramide synthase [45, 47] or/and sphingosine-1-phosphate, inhibiting sphingosine production [44]. These results suggest only certain degree of selectivity of tricyclic antidepressants. More experiments are needed to clarify this point.

3. Tricyclic Antidepressants Are Modulators of Cell Sensitivity to Insulin Action

It is known that sphingolipids play important role in the development of insulin resistance. A majority of studies evaluating the mechanisms of sphingolipid-induced insulin resistance have demonstrated that shingolipid, ceramide, is the most potent inhibitor of various parts in insulin signal transduction in target tissues. Ceramide is a product of hydrolysis of sphingomyelin by different sphingomyelinases or synthesis de novo. Ceramide inhibits glucose uptake or glycogen synthesis via inhibition of protein kinase Akt/PKB [12], blocks the translocation of Akt/PKB to the plasma membranes [14] and stabilizes interactions between Akt/PKB and atypical protein kinase C (PKC) isoform PKCζ [48, 49]. Ceramide, activating the protein phosphatase 2A (PP2A), could dephosphorylate Akt/PKB and prevent stimulation of glucose uptake by insulin [50]. The PP2A inhibitor, okadaic acid, added to the cultures of C2C12 myotubes [51], PC12 neurons [52], adipocytes [53], or human glioblastoma cell line [54] prevented the ceramide-induced disruption of insulin signaling pathway. In addition, ceramide could inhibit glucose uptake via Akt/PKB-independent pathway, attenuating glucose transporter 4 (GLUT4) translocation to cellular membranes [55]. Moreover, ceramide could decrease GLUT4 transcription and induce the insulin resistance in 3T3-L1 adipocytes [56].

Recently it has been demonstrated that ceramide, accumulated in the insulin responsive rat tissues, mediates, at least in part, the age-dependent and saturated fat-induced insulin resistance development. Long-term treatment of rats with dietetic saturated fat significantly increased the ceramide content and ceramide/sphingomyein ratio in the liver, muscles and brain cortex with respect to control ones [57]. These changes of ceramide levels in insulin responsive tissues were followed by reduction of insulin-stimulated glucose metabolism. Taking into account that the altered ceramide/sphingomyelin ratio may itself be a consequence of the elevated sphingomyelinase activity, our findings suggest that saturated fat, chronically added to the diet of rats, could increase the sphingomyelinase activity in the liver, muscles and brain. Moreover, sphingomyelinase-dependent ceramide accumulation could be, at least in part, the important reason of abnormal insulin signaling in the target cells. Actually, the long-term saturated fat feeding of the mice led to the pronounced increase of sphingomyelinases expression and activity in the liver and the epididymal and subcutaneous fat pads [58, 59]. High fat diet induced the expression of both neutral and acid sphingomyelinases in the fat tissues [58], increased activity of neutral sphingomyeinase but not acid sphingomyelinase in the liver [60] and activated acid sphingomyelinase in the colon [61].

Dietetic saturated fat-induced ceramide accumulation in the blood plasma and insulin responsive tissues is associated with the altered glucose homeostasis and development of the insulin resistance. Long-term saturated fat-enriched diet led to a sustained increase of glucose and insulin levels in the fasted state in the blood plasma [41, 42, 58]. Insufficient insulin-mediated suppression of plasma glucose after insulin administration to saturated fat-treated mice has been found. The decrease of high fat-induced expression of serine palmitoyltransferase, neutral and acid sphingomyelinase and ceramide content in the blood plasma and adipose tissue in plasminogen activator inhibitor-1$^{-/-}$ mice ameliorate insulin resistance [58]. Treatment of mice with an acid sphingomyelinase inhibitor, amitriptyline, significantly attenuated the diet-induced plasma ceramide level, improved glucose homeostasis and prevented development of the insulin resistance [41]. However, inhibitor of ceramide synthesis de novo, myriocin [62], administered to the saturated fat-treated Sprague-Dawley rats or Zucker diabetic fatty rats decreased the ceramide content in the blood serum, soleus muscle and liver and improved glucose homeostasis [15]. Myriocin reduced palmitic acid-induced ceramide accumulation in the liver cells, diaphragm muscle and neocortex, and abolished tissues resistance to insulin action of adult rats [57]. However, a specific inhibitor of key enzyme of sphingolipid synthesis, myriocin, or a specific inhibitor of the key enzyme of ceramide synthesis, fumonisin B1 [63], inhibited ceramide and sphingomyelin synthesis in the presence of [^{14}C]palmitic acid [64] and decreased but did not normalize elevated at old age ceramide content in the hepatocytes and only slightly increased cells sensitivity to insulin action [65, 66]. To elucidate the role of sphingomyelinases in changing the sphingomyelin and ceramide content in the liver cells of old rats, the effects of a functional inhibitor of acid sphingomyelinase, imipramine [17], and noncompetitive inhibitor of neutral sphingomyelinase, GW4869 [67], have been studied. We found that imipramine inhibits acid sphingomyelinase but does not affect the neutral sphingomyelinase in hepatocytes, while GW4869 inhibited neutral sphingomyelinase, but not the acid enzyme in the old liver cells [66]. Adding imipramine or GW4869 to the cell culture reduced the level of ceramide in hepatocytes of old rats, but not to the level of young animals. These results indicate that imipramine has a specific effect on the activity of acid sphingomyelinase and acid sphingomyelinase-dependent formation of ceramide in liver cells of old rats. Moreover, a sharp decline in the ability of old cells to respond to insulin could be partly improved by imipramine. Pretreatment of hepatocytes from old rats with imipramine or GW4869 had no significant effect on the basal level of glucose uptake and glycogen synthesis, but increased glucose metabolism in insulin-stimulated cells. However, only pretreatment of "old" liver cells with the mixture of inhibitors of ceramide synthesis and acid and neutral sphingomyelinases normalized ceramide content in the hepatocytes and improved cell sensitivity to insulin action. From these results it follows that hepatocytes resistance to insulin action depends on ceramide accumulation in "old" cells and that sphingomyelinases, as well as key enzymes of ceramide synthesis, are involved in the age-dependent ceramide oversupply and development of cells resistance to insulin action at old age.

More pronounced effect of tricyclic antidepressants on ceramide content and cell sensitivity to insulin action was demonstrated on another insulin classic target tissue, skeletal muscles. Two-week-long melipramin (imipramine) treatment reduced significantly the elevated ceramide content at old age and increased the sphingomyelin level in the extensor digitorum longus muscles [68]. In addition, melipramin decreased the ceramide/ sphingomyelin ratio, elevated at old age in muscle tissue, close to that in the adult animals. Melipramin administration to old rat significantly increased the extensor digitorum longus muscle sensitivity to insulin action. It has been shown that the drug increased the insulin-induced 2-dexy-D-[1,2- ^{3}H]glucose uptake by muscle tissue of 24-month-old rats. Moreover, using the diaphragm muscle of old rats it has been found that imipramine significantly decreased the acid sphingomyelinase activity (Figure 1) (unpublished data). The drug decreased the [N-methyl-^{14}C-phosphorylcholine]sphingomyelin hydrolyses and ^{14}C-phosphorylcholine release in the tissue lysates.

The 14-day-long imipramine treatment of 24-month-old rats involves a sustained decrease in ceramide content (Figure 2A) and ceramide/sphingomyelin ratio (Figure 2C) and increase in sphingomyelin level in the diaphragm of 24-month-old rats (Figure 2B) to the levels approaching those in adult 3-month-old animals. It is worth to note that imipramine nullified the age-dependent insulin resistance of diaphragm muscle. Treatment of old rats with imipramine caused a sustained increase of insulin-stimulated [3H]glucose uptake by diaphragm strips (Figure 3).

To establish whether the development of ceramide-dependent insulin resistance in the target tissues really depends on acid sphingomyelinase overexpression and to exclude the impact of ceramidase in insulin signaling, we additionally studied the effect of specific inhibitor of acid sphingomyelinase, zoledronic acid, on liver and muscle tissues (unpublished data). Zoledronic acid, which is a well-known drug against osteoporosis, showed a marked inhibition of acid sphingomyelinase with an IC 50 value of 5μM [69].

As expected, the zoledronic acid, as well as melipramine decreased the acid sphingomyelinase activity (Figure 1), ceramide content and ceramide/sphingomyelin ratio (Figure 2A, C) and increased the sphingomyelin content (Figure 2B) in the diaphragm of 24-month-old rats. In addition, 10-day-long treatment of old rats by zoledronic acid was accompanied by increase in tissue sensitivity to insulin action (Figure 3).

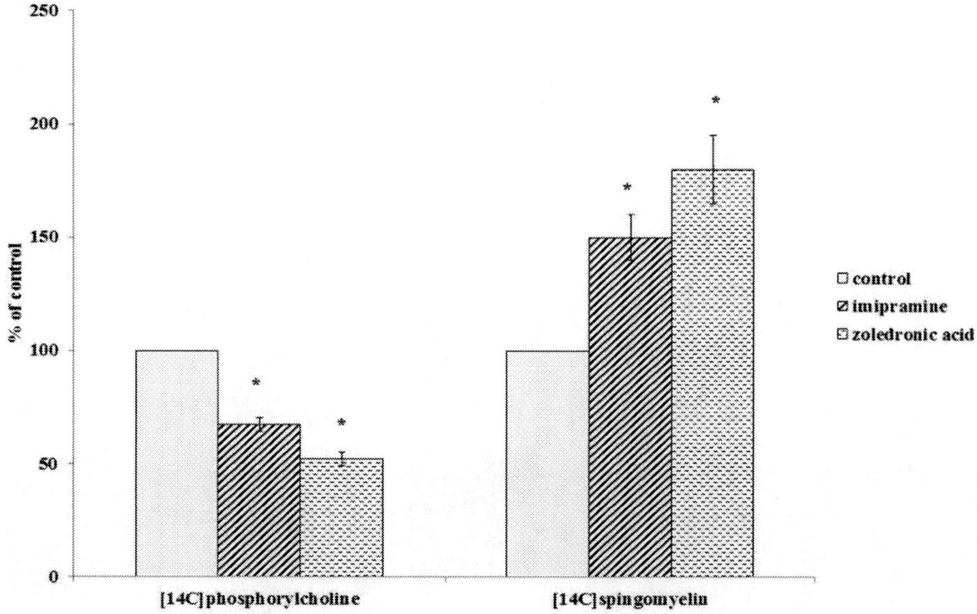

Figure 1. Imipramine and zoledronic acid inhibit the acid sphingomyelinase activity in the muscle tissues of old rats. Rats were intramuscularly injected with an imipramine (10 mg/kg body weight) daily for 14 days or injected with a zoledronic acid (0,15 mg/kg body weight) for 10 days, one time in two days. Control 24-month-old animals were intramuscularly injected with 0.9% NaCl for 14 days or 10 days. To determine the enzyme activity [methyl-^{14}C-phosphorylcholine]sphingomyelin was used as an enzyme subsrate. [^{14}C]phosphorylcholine release and non-hydrolyzed [^{14}C]sphingomyelin content were determined as described previously [71]. * $p<0.05$ drug-treated vs. control.

Addition of zoledronic acid, as well as imipramine, to the culture media induced the similar changes of the sphingolipid turnover and insulin-stimulated glucose uptake by muscle tissue (data not presented). These data avoided the possibility of antidepressant central action and demonstrated direct effect of imipramine on acid sphingomyelinase and thereby the skeletal muscle cells sensitivity to insulin action. In contrast, melipramine treatment of 24-month-old rats reduced sphingomyelinase activity [70], but did not change the ceramide level in the hippocampus and neocortex [71], while ceramide accumulation plays the central role in the development of brain age-dependent insulin resistance. Taking into account these results and data on normalization of ceramide content in brain structures by means of precursor of neutral sphingomyelinase inhibitor, gluthathione [72], reasonable assumption can be made, that development of insulin resistance in the brain at old age depends on activation of neutral, but not acid sphingomyelinase.

It is well documented that in skeletal muscle insulin resistance correlates highly with triacylglycerol content. However, the triacylglycerol in itself does not disrupt the insulin signaling in the target tissues. It is of interest, that diacylglycerol accumulation activated the PKC, phosphorylation of insulin receptor substrate-1 (IRS-1), which suppresses IRS-1 tyrosine phosphorylation and diminishes phosphatidylinositol 3-kinase (PI3K) activation and thus blocks the insulin signaling in the cells [73-77]. Free fatty acids accumulation in cells could cause not only the triacylglycerol and diacylglycerol oversupply, but the overproduction of known antagonist of PI3K-dependent signaling pathways, ceramide.

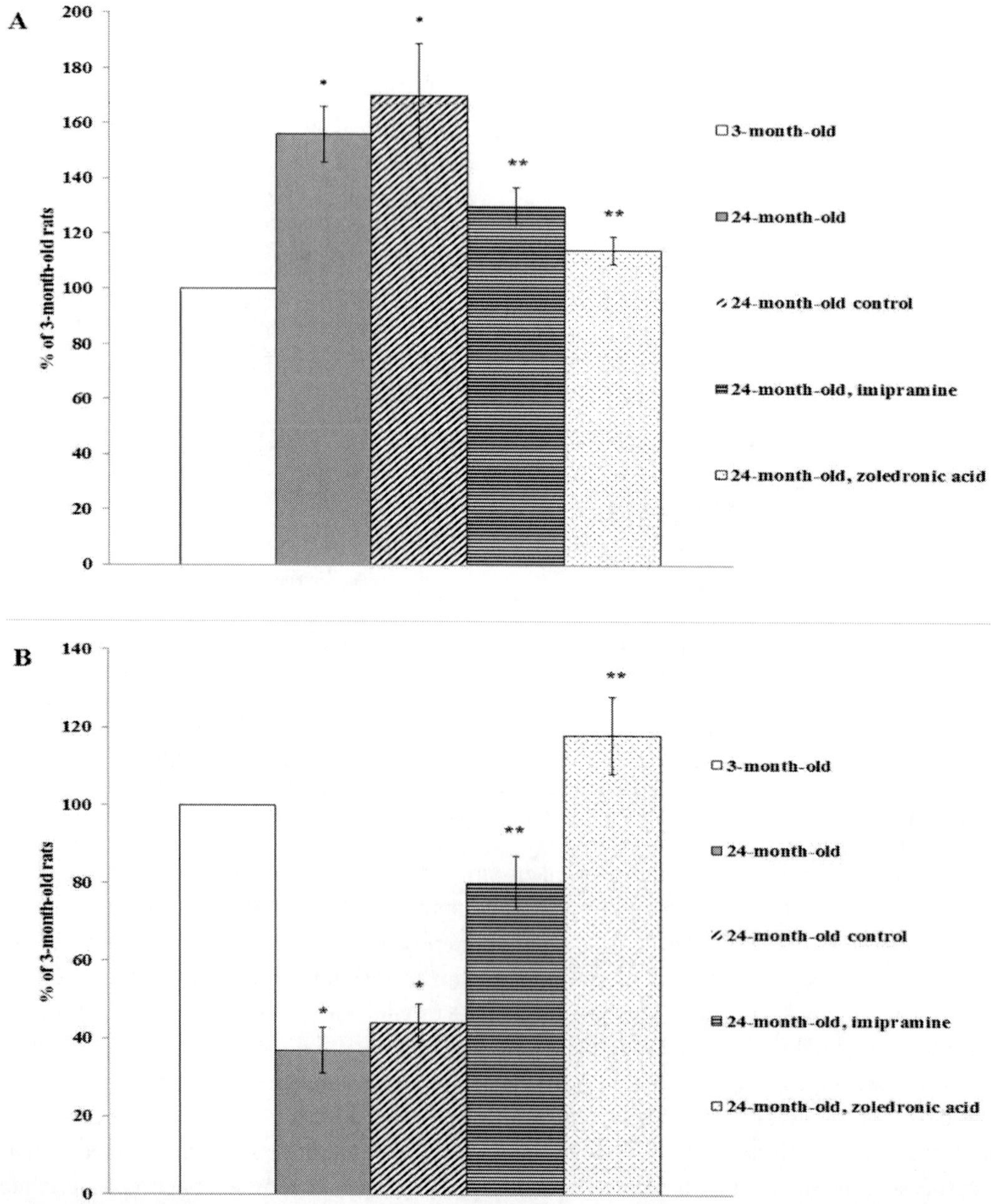

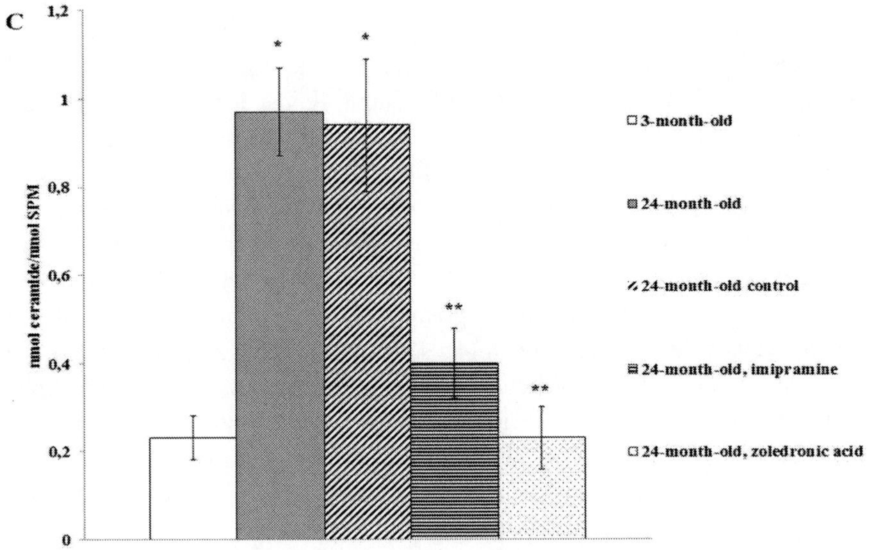

Figure 2. Effect of imipramine and zoledronic acid on sphingolipids content of the diaphragm of old rats. Sphingomyelin (SPM). A – ceramide; B – sphingomyelin; C – ceramide/sphingomyelin. 24-month-old rats were intramuscularly injected with an imipramine (10 mg/kg body weight) daily for 14 days or injected with a zoledronic acid (0,15 mg/kg body weight) for 10 days, one time in two days. Control 24-month-old animals were intramuscularly injected with 0.9% NaCl for 14 days or 10 days. * $p<0.05$ 24-month-old vs. 3-month-old rats; ** $p<0.05$ drug-treated vs. control.

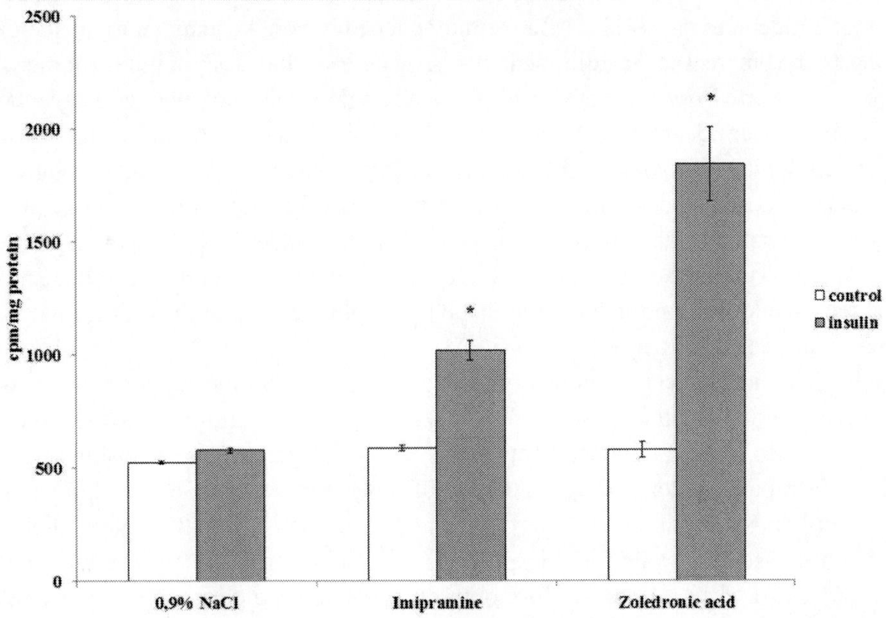

Figure 3. Imipramine and zoledronic acid abolish the age-dependent insulin resistance of muscle tissues. Rats were intramuscularly injected with an imipramine (10 mg/kg body weight) daily for 14 days or injected with a zoledronic acid (0.15 mg/kg body weight) for 10 days, one time in two days. Control 24-month-old animals were intramuscularly injected with 0.9% NaCl for 14 days or 10 days. The uptake of 2-D-[^{3}H]glucose by diaphragm strips were studied as described in [66]. * $p<0.05$ insulin-treated vs. control.

Culturing of the liver cells and diaphragm strips in the presence of palmitic acid increased significantly the contents of triacylglycerol, diacylglycerol, free fatty acids and ceramide and decreased stimulation of the [^{3}H]-D-glucose intake by insulin in the studied tissues [57, 65]. These findings suggest that accumulation of neutral lipids and ceramide in the insulin-responsive tissues is sufficient to inhibit the insulin-induced glucose uptake by the cells.

Melipramin administration to old rat significantly increased diaphragm strips sensitivity to insulin action (Figure 3) and normalized the lipids contents in soleus muscle [68]. Melipramin reduced the elevated triacylglycerol, diacylglycerol and free fatty acids content in diaphragm, gastrocnemius, soleus as well as neutral lipids in the extensor digitorum longus muscles. Taking into consideration that imipramine decreases the ceramide content and acid sphingomyelinase activity in the insulin responsive tissues at old age [68] and that pharmacological inhibition or genetic depletion of genes, controlling sphingolipid metabolism, prevents an excessive accumulation of neutral lipids in the liver and hepatic steatosis development [78], the assumption as to the possible use of the tricyclic antidepressants for the improvement of lipogenesis and target cell sensitivity to insulin action can be made.

However, antidepressant drugs-induced weight gain is a well known problem in the treatment of psychiatric disorders. The most depressed people that used the tricyclic antidepressants have abdominal adiposity, dyslipidemia and metabolic syndrome [79], while in some patients depression by itself is associated with weight gain [80]. Impaired glucose tolerance is usually observed in depressed patients, having an increased risk of diabetes mellitus development. The beneficial effects of antidepressants as well as opposite effects on glucose homeostasis have been tested in depressed and diabetic patients [81]. Some serotonergic antidepressants, such as fluoxetine, reduced hyperglycaemia, normalised glucose homeostasis and increased insulin sensitivity, whereas the desipramine (noradrenergic antidepressant) exerted opposite effects [82]. At the same time, duloxetine and venlafaxine (dual-mechanism antidepressants) did not disrupt glucose homeostatic dynamics. Nonselective inhibitor of hydrazine monoamine oxidase, phenelzine, induced hypoglycaemia and increased glucose disposal rate. The long-term (14-weeks-long) tricyclic antidepressants, amitriptyline or trimipramine, treatment of rats demonstrated the significant alterations in glucose homeostasis and some changes associated with insulin secretion and degradation [83]. The reasonable assumption has been made that prolonged use of these drugs may lead to insulin resistance and diabetes.

Weight gain side-effects of antidepressant drugs administration are strongly associated with increased lipid biosynthesis and dyslipidemia. Imipramine, amitriptyline and some other antidepressants, lead to a pronounced activation of sterol regulatory element-binding protein (SREBP) transcription factor and up regulation of downstream genes involved in cholesterol and fatty acid biosynthesis in cultured glial and liver cells [84, 85]. Nevertheless, the imipramine decreased the weight of mice after a four week-long treatment as compared to the controls [86] and had no effect on 24-month-old rat body weights after a two-week-long treatment [87]. It has been also demonstrated that imipramine abolished neutral lipids and free fatty acids accumulation in muscle tissues of old animals. Previously we reported that two-week-long imipramine administration to 24-month-old rats increased significantly the phosphatidylcholine and phosphatidylethanolamine contents in the blood serum, diaphragm, gastrocnemius and soleus muscles [88]. The tricyclic antidepressants administration resulted in decreased degradation and increased glycerolphospholipids synthesis in cells [89, 90]. It can not be excluded that the free fatty acids, newly synthesized in the liver, may be usable for glycerolphospholipids synthesis in the muscles rather than for triacylglycerol, diacylglycerol and ceramide synthesis. But, the mechanism of imipramine targeting of free fatty acids to glycerolphospholipids synthesis requires further investigations.

Tricyclic antidepressants can prevent lipids accumulation in the liver and steatosis as well as development of insulin resistance, induced by chronic ethanol feeding [91]. Ceramide administration to rats reproduced many features of insulin resistance associated with ethanol administration [92, 93]. Acid sphingomyelinase inhibitor, amitriptyline, reduced production of ceramide and development of steatosis, fibrosis, and nuclear pyknosis in the liver, and thereby increased the LEC rats (a genetic model for Wilson disease) survival [94]. Imipramine treatment of ethanol-fed mice blocks the ethanol-induced acid sphingomyelinase activation, ceramide generation, and ameliorates hepatic steatosis and glucose intolerance, impaired insulin signaling, and activation of stress kinases [91]. Authors hypothesized, that tricyclic antidepressants, as inhibitors of acid sphingomyelinase, can be a therapeutic target to treat alcoholic steatosis and thus reduce the hepatic morbidity of heavy ethanol consumption. Moreover, successful treatment with either a selective serotonin reuptake inhibitor or a tricyclic antidepressant increases the sensitivity to insulin in nondiabetic depressed patients [94]. However, the only short-term and not long-term treatment with antidepressants improves glucose homeostasis in nondiabetic depressed patients. Selective serotonin reuptake inhibitor treatment may improve glycemic control in depressed patients suffering with diabetes mellitus type 2, whereas noradrenergic antidepressants and tricyclic antidepressants may deteriorate the glycemic control [95].

CONCLUSION

Sphingolipids play an important role in cell physiology Altered sphingolipid turnover is usually associated with several human diseases such as inflammation, neurodegeneration and diabetes. Deficiency of acid sphingomyelinase results in the development of Niemann-Pick disease, while the up-regulation of acid sphingomyelinase in the cells results in progression of several disorders, including atherosclerosis, insulin resistance, diabetes, Alzheimers's disease, cystic fibrosis and Wilson disease [96-98]. Acid sphingomyelinase is activated in response to stress stimuli and at old age and plays the central role in the regulation of ceramide production and metabolism of other sphingolipids and glycerolipids in the cells. Inhibition of sphingomyelinase-dependent ceramide production in the cells is extremely important to ameliorate stress consequences. The functional inhibitors of acid sphingomyelinase are the most attractive candidates for treatment of such consequences. The tricyclic antidepressants are minimally toxic and widely used to protect cells from stress and to improve the acid sphingomyelinase-dependent metabolic dysfunctions. Moreover, the tricyclic antidepressants are licensed for medical use in humans and never completely downregulate the acid sphingomyelinase in the cells. Thus, functional inhibitors of acid sphingomyelinase do not lead to the storage of sphingomyelin in the cells and to the development of Niemann-Pick-like phenotype. Experiments on cell lines and different animal models of insulin resistance demonstrated that functional inhibitors of acid sphingomyelinase can prevent neutral lipids accumulation and increase target cells sensitivity to insulin action. The both beneficial and opposite effects of antidepressants on lipid and glucose homeostasis have been tested in depressed and diabetic patients. The effect of tricyclic antidepressants on the regulation of glucose metabolism in humans depends on the type of drug, duration of treatment, and patient metabolic state. Up to now it is not clear if tricyclic antidepressants influence on glucose homeostasis via acid sphingomyelinase inhibition or via central mechanism in humans. More research is needed to clarify this point.

REFERENCES

[1] Hannun YA, Obeid LM. Principles of bioactive lipid signalling: lessons from sphingolipids. *Nat. Rev. Mol. Cell Biol.* 2008 Feb;9(2):139–50.

[2] Cutler RG, Mattson MP. Sphingomyelin and ceramide as regulators of development and lifespan. *Mech. Ageing Dev.* 2001 Jul 15;122(9):895–908.

[3] Yu ZF, Nikolova-Karakashian M, Zhou D, Cheng G, Schuchman EH, Mattson MP. Pivotal role for acidic sphingomyelinase in cerebral ischemia-induced ceramide and cytokine production, and neuronal apoptosis. *J. Mol. Neurosci. MN.* 2000 Oct;15(2):85–97.

[4] Zhang Y, Mattjus P, Schmid PC, Dong Z, Zhong S, Ma WY, et al. Involvement of the acid sphingomyelinase pathway in uva-induced apoptosis. *J. Biol. Chem.* 2001 Apr 13;276(15):11775–82.

[5] Gulbins E, Bissonnette R, Mahboubi A, Martin S, Nishloka W, Brunner T, et al. FAS-induced apoptosis is mediated via a ceramide-initiated RAS signaling pathway. *Immunity.* 1995 Jan 4;2(4):341–51.

[6] Chung HS, Park SR, Choi EK, Park H-J, Griffin RJ, Song CW, et al. Role of sphingomyelin-MAPKs pathway in heat-induced apoptosis. *Exp. Mol. Med.* 2003 Jun 30;35(3):181–8.

[7] Kolesnick RN. 1,2-Diacylglycerols but not phorbol esters stimulate sphingomyelin hydrolysis in GH3 pituitary cells. *J. Biol. Chem.* 1987 Dec 15;262(35):16759–62.

[8] Megha, London E. Ceramide selectively displaces cholesterol from ordered lipid domains (rafts): implications for lipid raft structure and function. *J. Biol. Chem.* 2004 Mar 12;279(11):9997–10004.

[9] Nyberg L, Duan RD, Axelson J, Nilsson A. Identification of an alkaline sphingomyelinase activity in human bile. *Biochim. Biophys. Acta.* 1996 Mar 29;1300(1):42–8.

[10] Liu B, Hassler DF, Smith GK, Weaver K, Hannun YA. Purification and characterization of a membrane bound neutral pH optimum magnesium-dependent and phosphatidylserine-stimulated sphingomyelinase from rat brain. *J. Biol. Chem.* 1998 Dec 18;273(51):34472–9.

[11] Osawa Y, Seki E, Kodama Y, Suetsugu A, Miura K, Adachi M, et al. Acid sphingomyelinase regulates glucose and lipid metabolism in hepatocytes through AKT activation and AMP-activated protein kinase suppression. *FASEB J.* 2011 Apr;25(4):1133–44.

[12] Holland WL, Summers SA. Sphingolipids, insulin resistance, and metabolic disease: new insights from in vivo manipulation of sphingolipid metabolism. *Endocr. Rev.* 2008 Jun;29(4):381–402.

[13] Wymann MP, Schneiter R. Lipid signalling in disease. *Nat. Rev. Mol Cell Biol.* 2008 Feb;9(2):162–76.

[14] Stratford S, DeWald DB, Summers SA. Ceramide dissociates 3'-phosphoinositide production from pleckstrin homology domain translocation. *Biochem. J.* 2001 Mar 1;354(Pt 2):359–68.

[15] Holland WL, Brozinick JT, Wang L-P, Hawkins ED, Sargent KM, Liu Y, et al. Inhibition of ceramide synthesis ameliorates glucocorticoid-, saturated-fat-, and obesity-induced insulin resistance. *Cell Metab.* 2007 Jul 3;5(3):167–79.

[16] Summers SA, Nelson DH. A role for sphingolipids in producing the common features of type 2 diabetes, metabolic syndrome X, and Cushing's syndrome. *Diabetes.* 2005 Mar;54(3):591–602.

[17] Kornhuber J, Tripal P, Reichel M, Mühle C, Rhein C, Muehlbacher M, et al. Functional Inhibitors of Acid Sphingomyelinase (FIASMAs): a novel pharmacological group of drugs with broad clinical applications. *Cell Physiol. Biochem. Int. J. Exp Cell Physiol. Biochem. Pharmacol.* 2010;26(1):9–20.

[18] Arenz C. Small molecule inhibitors of acid sphingomyelinase. *Cell Physiol. Biochem. Int. J. Exp Cell Physiol. Biochem. Pharmacol.* 2010;26(1):1–8.

[19] Canals D, Perry DM, Jenkins RW, Hannun YA. Drug targeting of sphingolipid metabolism: sphingomyelinases and ceramidases. *Br. J. Pharmacol.* 2011 Jun;163(4):694–712.

[20] Gulbins E, Palmada M, Reichel M, Lüth A, Böhmer C, Amato D, et al. Acid sphingomyelinase-ceramide system mediates effects of antidepressant drugs. *Nat. Med.* 2013 Jul;19(7):934–8.

[21] Kornhuber J, Reichel M, Tripal P, Groemer TW, Henkel AW, Mühle C, et al. The role of ceramide in major depressive disorder. *Eur. Arch. Psychiatry Clin. Neurosci.* 2009 Nov;259 Suppl 2:S199–204.

[22] Kornhuber J, Müller CP, Becker KA, Reichel M, Gulbins E. The ceramide system as a novel antidepressant target. *Trends Pharmacol. Sci.* 2014 Jun;35(6):293–304.

[23] Tardy C, Codogno P, Autefage H, Levade T, Andrieu-Abadie N. Lysosomes and lysosomal proteins in cancer cell death (new players of an old struggle). *Biochim. Biophys. Acta.* 2006 Apr;1765(2):101–25.

[24] Schissel SL, Schuchman EH, Williams KJ, Tabas I. Zn2+-stimulated sphingomyelinase is secreted by many cell types and is a product of the acid sphingomyelinase gene. *J. Biol. Chem.* 1996 Aug 2;271(31):18431–6.

[25] Jin S, Yi F, Zhang F, Poklis JL, Li P-L. Lysosomal targeting and trafficking of acid sphingomyelinase to lipid raft platforms in coronary endothelial cells. *Arterioscler. Thromb. Vasc. Biol.* 2008 Nov;28(11):2056–62.

[26] Hannun YA, Luberto C. Ceramide in the eukaryotic stress response. *Trends Cell Biol.* 2000 Feb;10(2):73–80.

[27] Charruyer A, Grazide S, Bezombes C, Müller S, Laurent G, Jaffrézou J-P. UV-C light induces raft-associated acid sphingomyelinase and JNK activation and translocation independently on a nuclear signal. *J. Biol. Chem.* 2005 May 13;280(19):19196–204.

[28] Castillo SS, Levy M, Thaikoottathil JV, Goldkorn T. Reactive nitrogen and oxygen species activate different sphingomyelinases to induce apoptosis in airway epithelial cells. *Exp. Cell Res.* 2007 Jul 15;313(12):2680–6.

[29] Santana P, Pena LA, Haimovitz-Friedman A, Martin S, Green D, McLoughlin M, et al. Acid sphingomyelinase-deficient human lymphoblasts and mice are defective in radiation-induced apoptosis. *Cell.* 1996 Jul 26;86(2):189–99.

[30] Morita Y, Perez GI, Paris F, Miranda SR, Ehleiter D, Haimovitz-Friedman A, et al. Oocyte apoptosis is suppressed by disruption of the acid sphingomyelinase gene or by sphingosine-1-phosphate therapy. *Nat. Med.* 2000 Oct;6(10):1109–14.

[31] Pena LA, Fuks Z, Kolesnick RN. Radiation-induced apoptosis of endothelial cells in the murine central nervous system: protection by fibroblast growth factor and sphingomyelinase deficiency. *Cancer Res.* 2000 Jan 15;60(2):321–7.

[32] Lin T, Genestier L, Pinkoski MJ, Castro A, Nicholas S, Mogil R, et al. Role of acidic sphingomyelinase in Fas/CD95-mediated cell death. *J. Biol. Chem.* 2000 Mar 24;275(12):8657–63.

[33] Haimovitz-Friedman A, Cordon-Cardo C, Bayoumy S, Garzotto M, McLoughlin M, Gallily R, et al. Lipopolysaccharide induces disseminated endothelial apoptosis requiring ceramide generation. *J Exp Med.* 1997 Dec 1;186(11):1831–41.

[34] Lozano J, Morales A, Cremesti A, Fuks Z, Tilly JL, Schuchman E, et al. Niemann-Pick Disease versus acid sphingomyelinase deficiency. *Cell Death Differ.* 2001 Jan;8(1):100–3.

[35] Paris F, Fuks Z, Kang A, Capodieci P, Juan G, Ehleiter D, et al. Endothelial apoptosis as the primary lesion initiating intestinal radiation damage in mice. *Science.* 2001 Jul 13;293(5528):293–7.

[36] Separovic D, Pink JJ, Oleinick NA, Kester M, Boothman DA, McLoughlin M, et al. Niemann-Pick human lymphoblasts are resistant to phthalocyanine 4-photodynamic therapy-induced apoptosis. *Biochem. Biophys. Res. Commun.* 1999 May 19;258(3):506–12.

[37] Llacuna L, Marí M, Garcia-Ruiz C, Fernandez-Checa JC, Morales A. Critical role of acidic sphingomyelinase in murine hepatic ischemia-reperfusion injury. *Hepatol. Baltim. Md.* 2006 Sep;44(3):561–72.

[38] Rebillard A, Rioux-Leclercq N, Muller C, Bellaud P, Jouan F, Meurette O, et al. Acid sphingomyelinase deficiency protects from cisplatin-induced gastrointestinal damage. *Oncogene.* 2008 Nov 20;27(51):6590–5.

[39] García-Ruiz C, Colell A, Marí M, Morales A, Calvo M, Enrich C, et al. Defective TNF-alpha-mediated hepatocellular apoptosis and liver damage in acidic sphingomyelinase knockout mice. *J. Clin. Invest.* 2003 Jan;111(2):197–208.

[40] Henkes LE, Sullivan BT, Lynch MP, Kolesnick R, Arsenault D, Puder M, et al. Acid sphingomyelinase involvement in tumor necrosis factor alpha-regulated vascular and steroid disruption during luteolysis in vivo. *Proc. Natl. Acad. Sci. USA.* 2008 Jun 3;105(22):7670–5.

[41] Boini KM, Zhang C, Xia M, Poklis JL, Li P-L. Role of sphingolipid mediator ceramide in obesity and renal injury in mice fed a high-fat diet. *J. Pharmacol. Exp. Ther.* 2010 Sep 1;334(3):839–46.

[42] Deevska GM, Rozenova KA, Giltiay NV, Chambers MA, White J, Boyanovsky BB, et al. Acid sphingomyelinase deficiency prevents diet-induced hepatic triacylglycerol

accumulation and hyperglycemia in mice. *J. Biol. Chem.* 2009 Mar 27;284(13):8359–68.

[43] Elojeimy S, Holman DH, Liu X, El-Zawahry A, Villani M, Cheng JC, et al. New insights on the use of desipramine as an inhibitor for acid ceramidase. *FEBS Lett.* 2006 Aug 21;580(19):4751–6.

[44] Zeidan YH, Pettus BJ, Elojeimy S, Taha T, Obeid LM, Kawamori T, et al. Acid ceramidase but not acid sphingomyelinase is required for tumor necrosis factor-alpha-induced PGE2 production. *J. Biol. Chem.* 2006 Aug 25;281(34):24695–703.

[45] Jin J, Hou Q, Mullen TD, Zeidan YH, Bielawski J, Kraveka JM, et al. Ceramide generated by sphingomyelin hydrolysis and the salvage pathway is involved in hypoxia/reoxygenation-induced Bax redistribution to mitochondria in NT-2 cells. *J. Biol. Chem.* 2008 Sep 26;283(39):26509–17.

[46] Kitatani K, Idkowiak-Baldys J, Hannun YA. The sphingolipid salvage pathway in ceramide metabolism and signaling. *Cell Signal.* 2008 Jun;20(6):1010–8.

[47] Zeidan YH, Hannun YA. Activation of acid sphingomyelinase by protein kinase C delta-mediated phosphorylation. *J. Biol. Chem.* 2007 Apr 13;282(15):11549–61.

[48] Hajduch E, Turban S, Le Liepvre X, Le Lay S, Lipina C, Dimopoulos N, et al. Targeting of PKC zeta and PKB to caveolin-enriched microdomains represents a crucial step underpinning the disruption in PKB-directed signalling by ceramide. *Biochem. J.* 2008 Mar 1;410(2):369–79.

[49] Fox TE, Houck KL, O'Neill SM, Nagarajan M, Stover TC, Pomianowski PT, et al. Ceramide recruits and activates protein kinase C zeta (PKC zeta) within structured membrane microdomains. *J. Biol. Chem.* 2007 Apr 27;282(17):12450–7.

[50] Ugi S, Imamura T, Maegawa H, Egawa K, Yoshizaki T, Shi K, et al. Protein phosphatase 2A negatively regulates insulin's metabolic signaling pathway by inhibiting Akt (protein kinase B) activity in 3T3-L1 adipocytes. *Mol. Cell Biol.* 2004 Oct;24(19):8778–89.

[51] Chavez JA, Knotts TA, Wang L-P, Li G, Dobrowsky RT, Florant GL, et al. A role for ceramide, but not diacylglycerol, in the antagonism of insulin signal transduction by saturated fatty acids. *J. Biol. Chem.* 2003 Mar 21;278(12):10297–303.

[52] Salinas M, López-Valdaliso R, Martín D, Alvarez A, Cuadrado A. Inhibition of PKB/Akt1 by C2-ceramide involves activation of ceramide-activated protein phosphatase in PC12 cells. *Mol. Cell Neurosci.* 2000 Feb;15(2):156–69.

[53] Teruel T, Hernandez R, Lorenzo M. Ceramide mediates insulin resistance by tumor necrosis factor-alpha in brown adipocytes by maintaining Akt in an inactive dephosphorylated state. *Diabetes.* 2001 Nov;50(11):2563–71.

[54] Zinda MJ, Vlahos CJ, Lai MT. Ceramide induces the dephosphorylation and inhibition of constitutively activated Akt in PTEN negative U87mg cells. *Biochem. Biophys. Res. Commun.* 2001 Feb 2;280(4):1107–15.

[55] JeBailey L, Wanono O, Niu W, Roessler J, Rudich A, Klip A. Ceramide- and oxidant-induced insulin resistance involve loss of insulin-dependent Rac-activation and actin remodeling in muscle cells. *Diabetes.* 2007 Feb;56(2):394–403.

[56] Long SD, Pekala PH. Lipid mediators of insulin resistance: ceramide signalling down-regulates GLUT4 gene transcription in 3T3-L1 adipocytes. *Biochem. J.* 1996 Oct 1;319(Pt 1):179–84.

[57] Babenko, NA. Effects of short- and long-term saturated fat-enriched diet on the ceramide and neutral lipids accumulation in the insulin responsive tissues of rats. In: Langella JP, editor. Saturated fats: metabolism, disease risks, and public awareness. N.Y: Nova Science Publisher's; 2011; 71-97.

[58] Shah C, Yang G, Lee I, Bielawski J, Hannun YA, Samad F. Protection from high fat diet-induced increase in ceramide in mice lacking plasminogen activator inhibitor 1. *J. Biol. Chem.* 2008 May 16;283(20):13538–48.

[59] Lyn-Cook LE, Lawton M, Tong M, Silbermann E, Longato L, Jiao P, et al. Hepatic ceramide may mediate brain insulin resistance and neurodegeneration in type 2 diabetes and non-alcoholic steatohepatitis. *J. Alzheimers Dis. JAD.* 2009 Apr;16(4):715–29.

[60] Chocian G, Chabowski A, Zendzian-Piotrowska M, Harasim E, Lukaszuk B, Gorski J. High fat diet induces ceramide and sphingomyelin formation in rat's liver nuclei. *Mol. Cell Biochem.* 2010 Jul;340(1-2):125–31.

[61] Cheng Y, Ohlsson L, Duan R-D. Psyllium and fat in diets differentially affect the activities and expressions of colonic sphingomyelinases and caspase in mice. *Br. J. Nutr.* 2004 May;91(5):715–23.

[62] Wadsworth JM, Clarke DJ, McMahon SA, Lowther JP, Beattie AE, Langridge-Smith PRR, et al. The chemical basis of serine palmitoyltransferase inhibition by myriocin. *J. Am. Chem. Soc.* 2013 Sep 25;135(38):14276–85.

[63] Wang E, Norred WP, Bacon CW, Riley RT, Merrill AH. Inhibition of sphingolipid biosynthesis by fumonisins. Implications for diseases associated with Fusarium moniliforme. *J. Biol. Chem.* 1991 Aug 5;266(22):14486–90.

[64] Babenko NA, Hassouneh LKM, Kharchenko VS, Garkavenko VV. Vitamin E prevents the age-dependent and palmitate-induced disturbances of sphingolipid turnover in liver cells. *Age.* 2012 Aug;34(4):905–15.

[65] Babenko NA, Kharchenko VS. Ceramides inhibit phospholipase D-dependent insulin signaling in liver cells of old rats. *Biochemistry (Moskow).* 2012 Feb;77(2):180–6.

[66] Babenko NA, Kharchenko VS. Effects of inhibitors of key enzymes of sphingolipid metabolism on insulin-induced glucose uptake and glycogen synthesis in liver cells of old rats. *Biochemistry (Moskow).* 2015 Jan;80(1):104–12.

[67] Luberto C, Hassler DF, Signorelli P, Okamoto Y, Sawai H, Boros E, et al. Inhibition of tumor necrosis factor-induced cell death in MCF7 by a novel inhibitor of neutral sphingomyelinase. *J. Biol. Chem.* 2002 Oct 25;277(43):41128–39.

[68] Babenko NA, Timofiychuk OA. The role of acid sphingomyelinase in insulin-induced glucose uptake violations by muscle tissue in old age. *Problems of endocrine pathology (article in ukrainian).* 2013;(3):75–81.

[69] Roth AG, Drescher D, Yang Y, Redmer S, Uhlig S, Arenz C. Potent and selective inhibition of acid sphingomyelinase by bisphosphonates. *Angew Chem Int. Ed. Engl.* 2009;48(41):7560–3.

[70] Garkavenko VV, Storozhenko GV, Krasnikova ON, Babenko NA. Correction of age-related disorders of sphingolipid content in rat tissues by acid sphingomyelinase inhibition. *Int. J. Physiol. Pathophysiol.* 2012;3(3):281–6.

[71] Babenko NA, Shakhova EG. Long-term food restriction prevents aging-associated sphingolipid turnover dysregulation in the brain. *Arch. Gerontol. Geriatr.* 2014 Jun;58(3):420–6.

[72] Babenko NA, Shakhova OG. Effect of an inhibitor of sphingomyelinases, n-acetylcysteine, on cognitive functions in old rats. *Neurophysiology*. 2014 Jun 24;46(2):180–2.

[73] Morino K, Petersen KF, Shulman GI. Molecular mechanisms of insulin resistance in humans and their potential links with mitochondrial dysfunction. *Diabetes*. 2006 Dec;55 Suppl 2:S9–15.

[74] Schmitz-Peiffer C, Biden TJ. Protein kinase C function in muscle, liver, and beta-cells and its therapeutic implications for type 2 diabetes. *Diabetes*. 2008 Jul;57(7):1774–83.

[75] Neschen S, Morino K, Hammond LE, Zhang D, Liu Z-X, Romanelli AJ, et al. Prevention of hepatic steatosis and hepatic insulin resistance in mitochondrial acyl-CoA:glycerol-sn-3-phosphate acyltransferase 1 knockout mice. *Cell Metab*. 2005 Jul;2(1):55–65.

[76] Nagle CA, An J, Shiota M, Torres TP, Cline GW, Liu Z-X, et al. Hepatic overexpression of glycerol-sn-3-phosphate acyltransferase 1 in rats causes insulin resistance. *J. Biol. Chem*. 2007 May 18;282(20):14807–15.

[77] Benoit SC, Kemp CJ, Elias CF, Abplanalp W, Herman JP, Migrenne S, et al. Palmitic acid mediates hypothalamic insulin resistance by altering PKC-theta subcellular localization in rodents. *J. Clin. Invest*. 2009 Sep;119(9):2577–89.

[78] Bikman BT, Summers SA. Sphingolipids and hepatic steatosis. *Adv. Exp. Med. Biol*. 2011;721:87–97.

[79] Van Reedt Dortland AKB, Giltay EJ, van Veen T, Zitman FG, Penninx BWJH. Metabolic syndrome abnormalities are associated with severity of anxiety and depression and with tricyclic antidepressant use. *Acta Psychiatr Scand*. 2010 Jul;122(1):30–9.

[80] Chokka P, Tancer M, Yeragani VK. Metabolic syndrome: relevance to antidepressant treatment. *J. Psychiatry Neurosci*. 2006 Nov;31(6):414.

[81] Hennings JM, Schaaf L, Fulda S. Glucose metabolism and antidepressant medication. *Curr. Pharm Des*. 2012;18(36):5900–19.

[82] McIntyre RS, Soczynska JK, Konarski JZ, Kennedy SH. The effect of antidepressants on glucose homeostasis and insulin sensitivity: synthesis and mechanisms. *Expert Opin. Drug Saf*. 2006 Jan;5(1):157–68.

[83] Chadwick W, Wilson G, van de Venter M, Oelofsen W, Roux S. Shifts in metabolic parameters surrounding glucose homoeostasis resulting from tricyclic antidepressant therapy: implications of insulin resistance? *J. Pharm. Pharmacol*. 2007 Jan;59(1):95–103.

[84] Raeder MB, Ferno J, Glambek M, Stansberg C, Steen VM. Antidepressant drugs activate SREBP and up-regulate cholesterol and fatty acid biosynthesis in human glial cells. *Neurosci. Lett*. 2006 Mar 13;395(3):185–90.

[85] Raeder MB, Ferno J, Vik-Mo AO, Steen VM. SREBP Activation by antipsychotic- and antidepressant-drugs in cultured human liver cells: relevance for metabolic side-effects? *Mol. Cell Biochem*. 2006 May 23;289(1-2):167–73.

[86] Oriaifo SE, Omogbai EK. Comparative effects of imipramine, sertraline, nifedipine, furosemide and bumetanide on ingestive behaviour in mice. *Afr. J. Biotechnol*. 2013 Dec 11;10(78):18022–6.

[87] Babenko NA, Tymofiychuk O. Imipramine prevents lipid accumulation in skeletal muscle at old age. *WebmedCentral BIOCHEMISTRY* 2013;4(10):WMC004423 doi: 10.9754/journal.wmc.2013.004423

[88] Timofiychuk OA. Phospholipid content modulation by melipramin in serum and skeletal muscle of old rats. *Odessa Med. J. (Ukrainian).* 2012; 6: 54-57,.

[89] Sawada H, Takami K, Asahi S. A toxicogenomic approach to drug-induced phospholipidosis: analysis of its induction mechanism and establishment of a novel in vitro screening system. *Toxicol. Sci. Off. J. Soc. Toxicol.* 2005 Feb;83(2):282–92.

[90] Reasor MJ, Kacew S. Drug-induced phospholipidosis: are there functional consequences? *Exp. Biol. Med. Maywood* NJ. 2001 Oct;226(9):825–30.

[91] Liangpunsakul S, Rahmini Y, Ross RA, Zhao Z, Xu Y, Crabb DW. Imipramine blocks ethanol-induced ASMase activation, ceramide generation, and PP2A activation, and ameliorates hepatic steatosis in ethanol-fed mice. *Am. J. Physiol. Gastrointest Liver Physiol.* 2012 Mar 1;302(5):G515–23.

[92] De la Monte SM, Tong M, Nguyen V, Setshedi M, Longato L, Wands JR. Ceramide-mediated insulin resistance and impairment of cognitive-motor functions. *J. Alzheimers Dis. JAD.* 2010;21(3):967–84.

[93] Lang PA, Schenck M, Nicolay JP, Becker JU, Kempe DS, Lupescu A, et al. Liver cell death and anemia in Wilson disease involve acid sphingomyelinase and ceramide. *Nat. Med.* 2007 Feb;13(2):164–70.

[94] Weber-Hamann B, Gilles M, Lederbogen F, Heuser I, Deuschle M. Improved insulin sensitivity in 80 nondiabetic patients with MDD after clinical remission in a double-blind, randomized trial of amitriptyline and paroxetine. *J. Clin. Psychiatry.* 2006 Dec;67(12):1856–61.

[95] Deuschle M. Effects of antidepressants on glucose metabolism and diabetes mellitus type 2 in adults. *Curr. Opin. Psychiatry.* 2013 Jan;26(1):60–5.

[96] Smith EL, Schuchman EH. The unexpected role of acid sphingomyelinase in cell death and the pathophysiology of common diseases. *FASEB J. Off. Publ. Fed. Am. Soc. Exp. Biol.* 2008 Oct;22(10):3419–31.

[97] Jenkins RW, Canals D, Hannun YA. Roles and regulation of secretory and lysosomal acid sphingomyelinase. *Cell Signal.* 2009 Jun;21(6):836–46.

[98] Schuchman EH. Acid sphingomyelinase, cell membranes and human disease: lessons from Niemann-Pick disease. *FEBS Lett.* 2010 May 3;584(9):1895–900.

In: Sphingolipids
Editor: Angel Catalá

ISBN: 978-1-63483-019-5

Chapter 3

SPHINGOLIPID VARIETY, BIOSYNTHESIS AND REGULATION

Diana Molino[1,*] ***and Jean-Denis Faure***[2,3]
[1]Ecole Normale Supérieure-PSL Research University,
Département de Chimie, Sorbonne Universités,
Paris, France
[2]INRA, Institut Jean-Pierre Bourgin, Saclay Plant Sciences,
Versailles, France
[3]Agro Paris Tech, Institut Jean-Pierre Bourgin, Saclay Plant Sciences,
Versailles, France

ABSTRACT

Sphingolipids constitute a ubiquitous class of lipids present in all eukaryotes as well as in several prokaryotes and even some viruses. Their structure is peculiar among other classes of lipids with a backbone made of a sphingoid base bearing an N-linked fatty acid. This structure, known as ceramide, is further modified by means of polar head substitutions. Head substitutions, fatty acids of different lengths and sphingoid bases variants generate a great assortment of sphingolipid isoforms (sphingolipidome), which is highly diversified between organisms, kingdoms and cell types. Sphingolipids are essential for organism development and cell growth since complex sphingolipids are major components of biological membranes and involved in membrane domain organization and receptor signaling. Sphingoid bases are instead involved in cell signaling, cell growth and stress responses. Several key enzymes of the biosynthetic pathway are highly conserved and finely regulated to generate key "structural orthologs" that seem to be shared among organisms. In this chapter, we describe step by step the sphingolipid biosynthetic pathway, and via a continuous parallel between plant, yeast and animal, we will review the sphingolipid functions in different organisms and cells.

* Corresponding author. dmolino@gmail.com.

1. Introduction

The assortment of sphingolipids (sphingolipidome) is highly diversified between organisms, kingdoms and cell types; however, several key enzymes of the biosynthetic pathway are highly conserved and several "structural orthologs" seem to be shared among organisms (Figure 1). The sphingolipid family gathers ceramides and complex sphingolipids derived from ceramide by addition of a variety of head substitutions on the first carbon (C1). For example, in the model plant Arabidopsis, the substitution of glucose or glucosylinositol-phosphate leads to the formation of glucosyl-ceramide (GluCer) and glucosyl-inositol-phosphoryl-ceramide (GIPC), respectively. GluCer and GIPC represent around 40 and 50% respectively of the sphingolipidome of Arabidopsis seedlings (Markham, Li et al. 2006). GluCer is absent in *Saccharomyces cervisiae,* which only produces Inositolphosphoryl-ceramide (IPC) and Mannosylated Inositolphosphoryl ceramide (MIPC) (Lester and Dickson 2001). However, homology search using BLAST revealed that several other fungi species contain GluCer synthase enzymes. In animal cells, GluCer is also synthetized, but rapidly converted in more complex glycosphingolipids.

Sphingomyelin (SM) is another abundant sphingolipid found only in animal cell membranes. SM is characterized by the presence of phosphocholine as head group, instead of a carbohydrate molecule like in other sphingolipids. SM is greatly enriched in the multilayered myelin sheath surrounding the axons of neurons, where it participates in the propagation of the electrical impulses. It has been proposed that the intertwining hydrocarbon chains of SM strengthen myelin sheath packing.

Besides different head substitution, variability is also observed in FA chain length and LCB and/or FA saturation and hydroxylation, leading to a multiplicity of sphingolipid isoforms. In particular, hydroxylation of FA and LCB moieties is a hallmark of sphingolipids. Sphingolipids are the lipid with the highest level of hydroxylation in plants. Sphingolipidome, with more than 160 different molecules, appears to be more diversified in plants compared to animals and yeast (Markham and Jaworski 2007).

2. Sphingolipid Biosynthesis

2.1. The Biosynthesis of the Fatty Acyl Chain

Sphingolipids, as many other lipids, contain a fatty acyl chain. Fatty acids (FA) of 16 or 18 carbon atoms (C16, C18) are synthetized by the fatty acid synthase complex and released for further use in the form of Acyl-CoAs. By definition, fatty acids of C16 or C18 are named LCFA for long chain fatty acid. The fatty acid synthase complex is localized in plant chloroplasts or animal mitochondria. LCFA-CoA can be further elongated to VLCFA-CoA (very long chain fatty acids) in the ER by the elongase complex. The elongase complex is composed of 4 enzymes that act in a sequential mode to add 2 carbons to the LCFA-CoA. Multiple cycles lead to VLCFA-CoA from 20C up to 26C (Oh, Toke et al. 1997; Gaigg, Neergaard et al. 2001) (Han, Gable et al. 2002). In Arabidopsis these enzymes are: 3-keto acyl-CoA synthase (KCS) (Joubes, Raffaele et al. 2008), 3-keto acyl-CoA reductase (KCR) (Beaudoin, Wu et al. 2009), 3-hydroxy acyl-CoA dehydratase (HCD/PASTICCINO2) (Bach,

Michaelson et al. 2008), and the enoyl-CoA reductase (ECR/CER10) (Zheng, Rowland et al. 2005). VLCFA-CoA can then be incorporated in membrane lipids (sphingolipids, phospholipids); in storage lipids (triacylglycerides) or extracellular protective lipid (waxes and suberin).

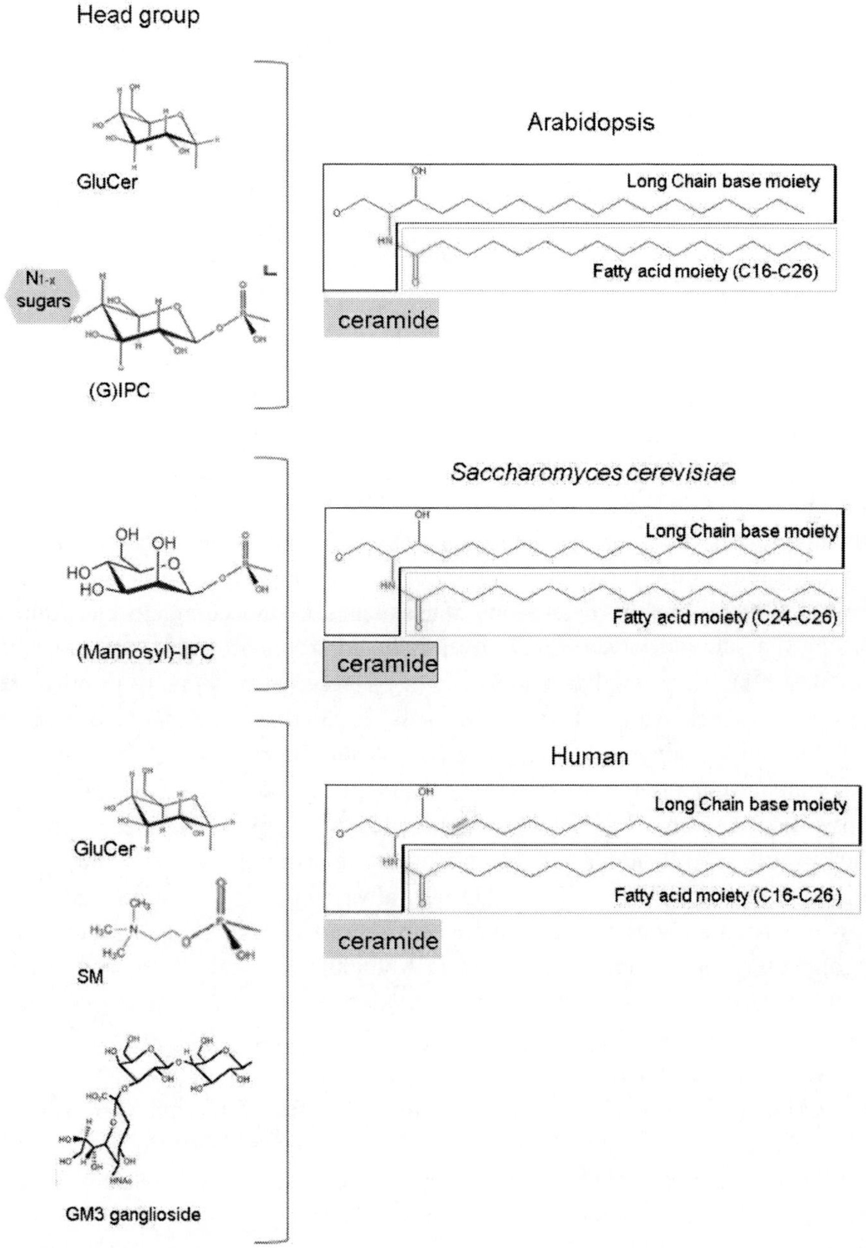

Figure 1. Sphingolipid structure and diversity. A common ceramide backbone is shared among the different organisms. Ceramide has two moieties: a Long Chain Base (LCB) and an N-linked fatty acid. Polar head groups (left) and fatty acyl chain length (right) are indicated for Arabidopsis, Saccharomyces cerevisiae and humans. Ceramide structure is from Pata et al. 2010.

The length of the FA chain is a critical feature of sphingolipids. In both animals and plants, it can be long (LCFA: C16-18), or very long (VLCFA: C20-26), while in yeast, sphingolipids only contain VLCFA (C24-26). Sphingolipid biosynthesis is therefore tightly associated with FA elongation in yeast, while in plants and animals, two branched pathways exist, one linked to FA elongation, and the other not. In both animals and plants, LCFA-C18 and VLCFA-C24 are both significantly abundant and they may fulfill different functions. Phospholipids may also contain VLCFA but the great majority of them contain two symmetric C16/C18 chains. Finally, most of the FAs associated with sphingolipids are hydroxylated and saturated or mono-unsaturated.

In summary sphingolipids are characterized by very long, saturated or monounsaturated and hydroxylated acyl chains.

2.2. Sphingoid Bases Variants: Hydroxylation Is Essential in Plants while Saturation Predominates in Animals

The first step of the biosynthesis of the sphingoid backbone is the condensation of a serine with palmitoyl-CoA, via a serine palmitoyl transferase (SPT) (Table 1, Figure 2). The product is a long chain base called 3-ketosphinganin, which is immediately reduced by the ketosphinganin reductase (KSR) to sphinganine (Table 1). Genes encoding SPT are present in a large number of species from bacteria to humans and highly conserved among eukaryotes (Sperling and Heinz 2003; Lynch and Dunn 2004). The importance of sphingolipids in cell biology is highlighted by the deleterious consequence of the complete knockout of *SPT*. Mutations in SPT subunits have been involved in hereditary neuropathy in humans (Dawkins, Hulme et al. 2001). The complete knockout of *SPT* resulted in lethality in mice, yeast and plants (Hojjati, Li et al. 2005; Chen, Han et al. 2006) revealing that LCBs are essential molecules for cell viability. SPT functions as a heterodimer, and a 20% reduction of its activity by the inactivation of one subunit is sufficient for impairing cell expansion in Arabidopsis (Chen, Han et al. 2006). A complete list of all enzymes involved in the biosynthesis and modifications of sphingolipids in Arabidopsis and their subcellular localization is listed in Table 1. The LCB moiety always contains 18 carbon atoms chain that can be hydroxylated and desaturated. Such chemical modifications add structural variability to sphingolipids. A total extract of LCBs from Arabidopsis reveals a majority of di- and tri-hydroxylated LCBs (d18 or t18). Di-hydroxylation is associated with the presence of hydroxyl groups on C1 and C3, corresponding to the serine and palmitoyl-CoA precursors, respectively, while the third hydroxylation occurs later in the pathway on C4 by a C4-hydroxylase (Figure 2). In Arabidopsis, two redundant genes, *SBH1* and *SBH2* (for sphingoid base hydroxylase 1 and 2, Table 1), encode a C4-hydroxylase. Double mutants as well as RNAi suppression lines revealed the physiological and metabolic relevance of this hydroxyl group (Chen, Markham et al. 2008). The complete loss of t18 (tri-hydroxyl LCB) resulted in both cell division and cell expansion defects, stressing the importance of LCB hydroxylation in sphingolipid cellular function. Moreover, the total content of sphingolipids in *sbh* mutants was globally modified with higher levels of d18 (di-hydroxyl LCB), as predicted. Unexpectedly, these mutants also showed higher levels of sphingolipids with shorter fatty acyl chains. These observations suggest that the 3-hydroxyl configuration (t18) is preferentially combined with VLCFA, while d18 is mainly associated with LCFA. Chen at al.

proposed that the presence of tri-hydroxyl groups would be an important structural determinant for channelling VLCFA downstream in complex sphingolipids (Chen, Markham et al. 2008). Indeed, in a total extract of *Arabidopsis* seedlings, most of the GluCer containing d18 is associated with C16 FA to form d18/C16-GluCer. It is not true for GIPC which overall contains mostly t18 LCB even in C16-GIPC. The d18 LCB are thus channeled in C16 GluCer while t18 are mostly associated with VLCFA- GluCer and GIPC. The fact that in *sbh* mutants d18/C16 GluCer and GIPC are increased 10-fold, support the role of C4-hydroxylation in channeling t18 species into VLCFA-complex sphingolipids.

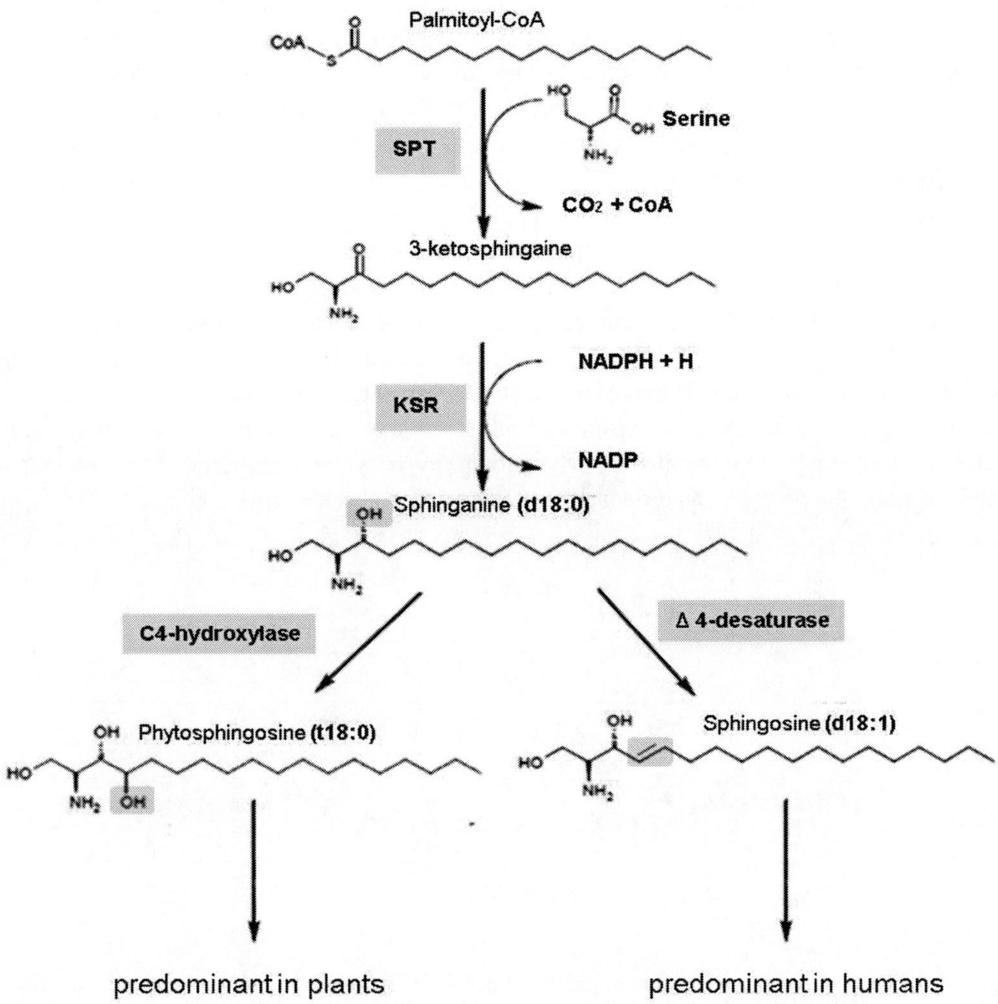

Figure 2. Sphingoid base biosynthesis. SPT for serine palmitoyl transferase. KCR for keto sphinganine reductase. d18:0, t18:0, d18:1 nomenclature indicates the number of -OH (di or tri) and the number of double bound. Predominant LCBs species in animals and plants are indicated.

LCBs can also be unsaturated with 1 or 2 double bonds on sphinganine to produce d18:1, d18:2 or on phytosphingosine to produce t18:1, or t18:2. Desaturation can occur at C4 in *trans* configuration or on C8 in both *cis* and *trans* configurations. The enzymes responsible for these desaturations are the Δ4- or Δ8-desaturases. Combination of these different LCB modifications (hydroxylation and desaturation) with their respective stereoisomers results in a subset of 9 different molecules (Figure 3). The Δ4-desaturase (DES4) was recently reported to be expressed in specific tissues, mainly pollen and at low level in flowers (Michaelson, Zauner et al. 2009). The complete knockout of the DES4 does not lead to any developmental defect. However the *des4* sphingolipidome of flower tissue revealed low levels of C16-GluCer, while VLCFA-GluCer and VLCFA-GIPC levels were unaffected, indicating that Δ4-desaturated LCBs are preferentially combined with C16 rather than with longer fatty acyl chains. Interestingly, mutation of the *Drosophila* LCB desaturase homolog DES1 caused defective spermatogenesis in flies confirming the importance of unsaturated LCB in development (Endo, Akiyama et al. 1996). It has to be underlined that the C4-hydroxylation and Δ4-desaturation occur on the same site, suggesting that the respective enzymes may act in a competitive manner for their substrate. Interestingly, while Δ4-desaturation does not seem to be essential in Arabidopsis, it is prevalent in animal sphingolipids, leading to the formation of sphingosine (SPH), an important signal molecule (see below) (Figure 2). On the other hand, Δ8-desaturation, either in *cis* or *trans* configuration, is very common in plant sphingolipids (Sperling, Franke et al. 2005). The proportion of LCB and their *cis/trans* isomers is highly variable between plant species. For example, 20% of the total GluCer in Arabidopsis leaves is $t18{:}1\Delta8^{E}$, while the $t18{:}1\Delta8^{Z}$ is only 3.4% of GluCer. In maize leaves, the levels of t:18:1 GluCer is low compared to Arabidopsis, furthermore the relative distribution of the *E/cis* and *Z/trans* are inverted with respectively 0.2% and 1% of total GluCer, reviewed by (Pata, Hannun et al. 2010). To date, the exact order in which C-4 hydroxylation, Δ4 and Δ8 desaturations occur is not yet fully defined, and the substrate specificities of each step remains poorly known. A possible sequence of the enzymatic activities along the pathway, as proposed by Pata et al. 2009 is shown in Figure 3.

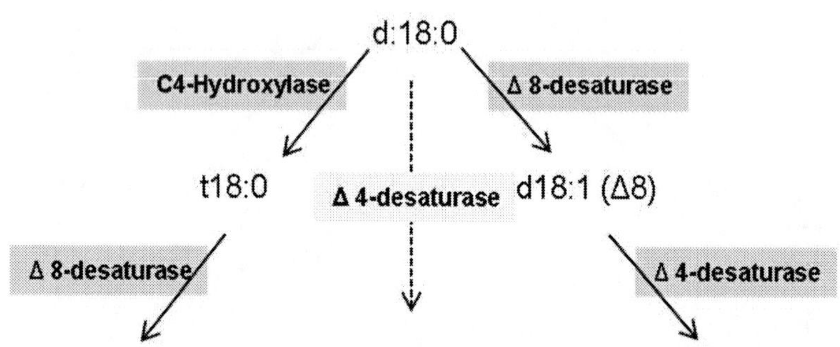

Figure 3. LCBs modifications. Possible sequential order of LCB modifications. The pathway does not take into account the different possible isomers.

2.3. Ceramide Synthesis and Inhibition, an Essential Component of Life Span

After LCB synthesis, the next step in sphingolipid biosynthetis involves the condensation of a LCB with a fatty acyl chain to form ceramide (Cer). The enzyme responsible for this reaction is the ceramide synthase (CerS). The CerS gene was originally isolated during a screen designed to identify genes that were differentially expressed cells that had undergone low or high number of divisions and that could thus be associated with longevity (Egilmez, Chen et al. 1989). Like most cells, yeast cells are subjected to aging, which is defined as the total number of budding events (cell divisions). A gene, whose expression was high in early generations, decreased linearly during aging until being undetectable at final age (around 15 divisions), was characterized and named LAG1, for longevity assurance gene (D'Mello N., Childress et al. 1994). *lag1* mutant displayed no defect in size, shape, budding pattern or rate of division; however, it exhibited increased life span corresponding to an average of 25 divisions (D'Mello N., Childress et al. 1994). Considering the negative correlation between LAG1 expression level and the increased life span, the author proposed that Lag1p could be directly or indirectly responsible for the average of 20-cell cycle limit in yeast life span. Another gene displaying a significant homology with LAG1 was found in yeast and named LAC1. The double knockout *lag1/lac1* was either unviable or had severe growth defects depending on the strain background (Barz and Walter 1999). Both proteins were localized in the ER and the viable double mutant combination displayed inability to transport GPI-anchored proteins from the ER to the Golgi, a defect associated with sphingolipid levels (Barz and Walter 1999). Indeed, LAG1 and LAC1 disruption resulted in reduced sphingolipid biosynthesis, because both genes were found to code for Acyl-CoA dependent ceramide synthases (Guillas, Kirchman et al. 2001; Schorling, Vallee et al. 2001).

Mammals, like humans or mice, have six different LAG1 homologs named LASS1-6 (for longevity assurance) that code for six different ceramide synthases (CerS1-6). The different enzymes display different substrate specificity, related in particular to the fatty acyl chain (Pewzner-Jung, Ben-Dor et al. 2006; Mizutani, Mitsutake et al. 2009) (Table 2). CerS3 is the only ceramide synthase utilizing a broad range of FA substrates, mainly C18 and C24, but also medium chains (C14, C16). Surprisingly, mammalian CerS, with the exception of CerS1, present an extra domain. This domain is a bona fide homeodomain except that it defines a very divergent subclass. (Venkataraman and Futerman 2002; Riebeling, Allegood et al. 2003; Holland, Booth et al. 2007). Although bona fide domain structure prediction studies suggest that they form three alpha helices in the correct conformation to bind DNA, no evidence for a DNA binding or transcription factor activity has been demonstrated yet (Pewzner-Jung, Ben-Dor et al. 2006). However, ceramide synthases are ER transmembrane proteins and predicted to be inside the ER lumen. It is therefore unlikely that they could be in contact with DNA. Yeast and plants LAGs protein do lack this homeodomain.

The high number of CerSs in humans likely reflects tissue specific functions. CerS have been mainly characterized in cell cultures and a knockout model is only available for CerS2 (Table 2) (Pewzner-Jung, Park et al. 2010). The cerS2$^{-/-}$ mice could survive almost 2 years (like the wild type) but exhibited important liver dysfunction. Specifically, hepatocytes cerS2$^{-/-}$ presented a shortened life span, with defective cellular regeneration that caused the development of liver carcinoma (Pewzner-Jung, Park et al. 2010). CerS3 (Table 2) was found to be highly expressed in the skin (Riebeling, Allegood et al. 2003), which was correlated to the involvement of VLCFA-ceramides in the impermeable barrier function of the skin (Coderch, Lopez et al. 2003). LCFA-ceramide, namely C16-ceramide (Table 2), is the most abundant sphingolipid species in non-neuronal tissues, and its accumulation has been associated, in several cases, with the induction of apoptosis and cell stress (Eto, Bennouna et al. 2003; Osawa, Uchinami et al. 2005; Senkal, Ponnusamy et al. 2011). Tumor cells such as head and neck squamous carcinoma, melanoma and Lewis lung carcinoma, also contained high levels of C16-Cer (Koybasi, Senkal et al. 2004) and the accumulation of such short species were associated with drug resistance and cell death (Ponnusamy, Meyers-Needham et al. 2010). However, length of fatty acyl chain of sphingolipids could have opposite effects on carcinoma depending wether it occurred in the lung and colon carcinoma (Hartmann, Lucks et al. 2012).

Our knowledge of the role of ceramide synthase activity was not only inferred from genetic studies, but also from the use of specific toxin-derived inhibitors. The fungus *Fusarium moniliforme* is a common contaminant of maize (*Zea Mays*) intended for human and animal consumption (Marasas, Kriek et al. 1984). The presence of *F. moniliforme* in maize has been correlated with high incidence of human oesophageal cancer in southern Africa and outbreaks of leukoencephalomalacia in horses (Pienaar, Kellerman et al. 1981; Rabie, Marasas et al. 1982; Gelderblom, Jaskiewicz et al. 1988). The toxic and carcinogenic contaminants were isolated as Fumonisins (made by fusarium moniliforme) by (Gelderblom, Jaskiewicz et al. 1988). Fumonisin B1 (FB1) was found to be the most abundant and active fumonisin, inducing liver toxicity and cancer after 1 month of treatment in rats (Gelderblom, Jaskiewicz et al. 1988). The hepatotoxicity was correlated with both apoptosis and necrosis. Additionally, FB1 caused the appearance of foci of cell proliferation and carcinogenesis (Gelderblom, Kriek et al. 1991; Gelderblom, Lebepe-Mazur et al. 2001). *In vitro* studies in rat hepatocytes revealed that FB1 targets sphingolipid biosynthesis by increasing LCB levels and reducing the ceramides and complex sphingolipids levels (Yoo, Norred et al. 1992; Merrill, van Echten et al. 1993). It was also shown that FB1 could directly inhibit the enzymatic activity of ceramide synthase *in vitro*, and the potency of the inhibition was correlated with the accumulation of free LCBs and acyl-chain substrates (Merrill, Wang et al. 1993). Elevated level of sphingosine in urine was therefore used as a diagnostic indicator of FB1 consumption and the toxicity associated with altered sphingolipid metabolism.

In Arabidopsis, 3 LAG1 homologs are present in the genome and were named LOH1-3, for LAG One Homolog). Ceramides represent only 1-2% of total sphingolipids extract, as they are rapidly converted into more complex forms. Two major classes of ceramides exist in the Arabidopsis sphingolipidome: ceramide and hydroxyl-ceramides (hCer), the latter containing α-hydroxylated FAs (Lynch and Dunn 2004; Markham and Jaworski 2007). α-hydroxylation is however mostly found in complex sphingolipids downstream of ceramide (see below), suggesting that either only α-hydroxy-Cer pool is used for downstream biosynthesis of complex sphingolipids or that α-hydroxylation occurs on a fraction of ceramide pool and independently on the whole pool of complex sphingolipids. The biological role of α-hydroxylation is unknown, but it is tempting to speculate that it plays a structural role in mediating interactions between lipids or between lipids and proteins within the membrane layer. Arabidopsis mutants depleted in very long acyl chain sphingolipids were non-viable contrary mutants lacking long acyl chain sphingolipids demonstrating the importance of the length of acyl chain (Markham, Molino et al. 2011). VLCFA-sphingolipids were in particular involved in sorting specific material from the TGN towards the PM *in vivo* (Markham, Molino et al. 2011). VLCFA-sphingolipid depletion modified the dynamic of specific endomembrane compartments in particular for vesicle fusion during exocytosis and cytokinesis (Molino, Van der Giessen et al. 2014).

2.4. Complex Sphingolipids in Plants

Downstream from ceramide synthase, several enzymes transfer a head group to the C1 of the ceramide backbone (Figure 1). Yeast, mammals and plants share a common head substitution but show also some specificity. In yeast, the IPC-synthase transfers a phosphorylinositol group from a phosphatidylinositol (PI) to ceramide to form Inositolphosphoryl-ceramide (IPC). In mammals, sphingomyelin synthases (SMS) transfer a phosphocholine from phosphatidylcholine to ceramide to produce sphingomuelin (SM), the most abundant mammalian sphingolipid. Apart from SM, mammals synthesize a neutral complex sphingolipid, glucosylceramide (GluCer) by transferring glucose to ceramide via glucosyceramide synthase (GCS). In plants, two major complex sphingolipids are found: a glucosylated ceramide (GluCer) and a glucosylated inositolphosphoryl-ceramide (GIPC). In Arabidopsis, GluCer-synthase (GCS) is encoded by a unique essential gene, and was recently shown to be important for Golgi morphology and protein secretion (Melser, Batailler et al. 2010).

Furthermore, there are three functional IPC-synthases (Mina, Okada et al. 2010). The first IPC-synthase identified was the gene product of ERH1 (enhancing RPW8-mediated HR-like cell death) (Wang, Yang et al. 2008). ERH1 was identified in a screening for mutations that enhanced hypersensitive response during pathogen infection (see below). The mutant *erh1* did not display an obvious reduction in GIPC, probably because of redundancy with the two other IPC synthases, however it accumulated ceramides which are considered as the main cause of the increased cell death responses in the mutant.

GIPC and GluCer can be distinguished by their head groups as well as by their fatty acid and LCB contents. A complete sphingolipid profile of Arabidosis provided a detailed picture of the diversity of sphingolipids among the four classes of molecules; Cer, α-hydroxy-Cer (containing α-hydroxylated fatty acid), GluCer and GIPC (Markham, Li et al. 2006). Four major observations can be made from these profiles. Firstly, t18 is the predominant form in both GluCer and GIPC, representing 90% of the total LCBs species. Secondly, t18 preferentially associates with VLCFA-Sphingolipids and d18 with LCFA-Sphingolipids. Thirdly, Cer can be α-hydroxylated or not (Cer versus hydroxy-Cer), while complex GluCer and GIPC only contain α-hydroxy FA. Finally, we have two major forms of GluCer: d18-LCFA(C16), t: 18-VLCFA(C24), and only one major form of GIPC: t:18 VLCFA(C24).

Sphingolipid distribution could also differ between different tissues suggesting specialized metabolic pathways. In plants, sphingolipids showed qualitative and quantitative variations according to the different tissues (Islam, Jacquemot et al. 2012; Tellier, Maia-Grondard et al. 2014). GIPCs and GluCer showed for instance quantitative variations between seedling and seed extracts however seeds but not oil showed specific accumulation of a new amine derived GIPC (Tellier, Maia-Grondard et al. 2014). The presence of specific sphingolipid species in specific tissues raises questions about their roles that need to be addressed.

3. Localisation of Sphingolipid Biosynthetic Enzymes and Sphingolipid Transport

Yeast IPCS, the human sphingomyelin synthase 1 (SMS1) and human GCS are all found in the Golgi (Levine, Wiggins et al. 2000; Huitema, van den Dikkenberg et al. 2004; Tafesse, Huitema et al. 2007), while the human SMS2 is at the plasma membrane (PM) (Yamaji and Hanada 2015). Interestingly in Arabidopsis, LOHs and GCS localize in the ER (Marion, Bach et al. 2008), IPCS-ERH1 has been instead localized at the trans Golgi network (TGN) (Wang, Yang et al. 2008). These different localizations indicate that the compartmentalization of the sphingolipid pathway and the channeling of ceramide into GluCer or IPC would depend on ceramide transport. Direct (non- vesicular) translocation of ceramide to the Golgi was proposed to occur thanks to ceramide transfer protein CERT (Cer transfer) (Hanada, Kumagai et al. 2003). CERT contains a ceramide binding domain (START) and a plekstrin (PH) domain, needed for binding to PtdIP (phosphatidylinositol-phospate) at the Golgi. It's integrity is essential for the synthesis of SM. Localization of a fluorescent Cer to the Golgi was also impaired in Hela cells expressing mutated CERT. CERT could also extract Cer from ER membranes *in vitro* demonstrating its role as ceramide carrier.

In Arabidopsis, a ceramide transfer protein has not been identified yet. However, a putative sphingosine-transfer protein has been isolated in a screening for mutants with accelerated cell death symptoms and named *acd11* (for accelerated cell death) (Brodersen, Petersen et al. 2002). ACD11 is part of the GLTP protein family (glycolipid transfer protein) and at least two other homologs are present in the Arabidopsis genome, but the precise biochemical activity of these proteins awaits further work (Table 1).

4. Sphingolipids Are Major Membrane Lipids

Lipids are a vast class of molecules, which cover different cellular functions such as energy storage, protective barriers, membranes biogenesis and dynamics, and intracellular signaling.

The majority of lipids that build up a biological membrane are amphiphilic, with a hydrophilic head and a lipophilic tail. The polar head interacts with water while the hydrophobic tails tend to self-associate to avoid water, leading to the spontaneous formation of a bilayer which maximizes these hydrophilic/hydrophobic interactions. The propensities of lipids to self-assemble in simple or more complex structures could be one of the major mean of determining the different shapes found within the cells. Lipids can be organized as a flat surface for plasma membrane, spherical and the tubular structures in the vesicular system, and more intricate geometry in ER (endoplasmic reticulum) and Golgi (Melser et al. 2010). Sphingolipids play major roles in biological membrane and participate in membrane organization. Specific and diverse PM structures such as plasmodesmata of plant tissues or immunological synapsis of animal cells from the immunity system have been shown to need sphingolipids for both structural and functional meaning (Gombos, Kiss et al. 2006; Grison, Brocard et al. 2015). Sphingolipids play essential roles in PM receptor clustering and signaling and malfunctioning of these processes have been associated with pathology. (Lauwers and Andre 2006; Lingwood and Simons 2010; Molino and Galli 2014).

The best way to look at the intracellular organization is to consider the possible routes followed by a plasma membrane (PM)-targeted protein after its synthesis in the ER. In a simplified view, after synthesis in the ER membranes a PM-protein usually reaches the Golgi apparatus. The Golgi is composed of a *cis* part which is in continuity with ER membranes, a medial, and a *trans* part which constitutively delivers proteins to all post-Golgi compartments, including the plasma membrane, the different endosome populations, and the prevacuolar and vacuolar compartments. Such a step by step transport defines the eukaryotic secretory pathway (Figure 4).

Table 1. Enzymes and genes from *Arabidopsis thaliana* involved in sphingolipid biosynthesis

ENZYME	Name	Gene	Localization	References
serine palmitoyl transferase (heterodimeric)	LCB1	At4g36480	ER	Chen et al. 2006 Tamura et al. 2001
	LCB2a	At5g23670	ER	
	LCB2b	At3g48780		Teng et al. 2008
ceramide synthase	LOH1	At3g25540	ER	Markham, Molino et al. 2010
	LOH2	At3g19260	ER	Markham, Molino et al. 2010
	LOH3	At1g13580	ER	Markham, Molino et al. 2010
inositol phosphoryl ceamide synthase	IPCS	At2g37940	Golgi	Wang et al. 2008
		At2g29525 At3g54020	? mitocondria	
glucosyl inositol phophoryl ceramide synthase?	GIPCS?	unknown	unknown	Mina et al. 2010
glucosyl ceramide synthase	GCS	At2g19880	ER	Melser et al. 2009
ceramidase	CDase	At4g22330	ER/Golgi	Pata et al. 2008
glucosyl ceramidase	GlcCDase	unknown	unknown	---
glycolipid transfer protein	GLTP1	At2g33470	ER/Golgi	West et al. 2008
glycolipid binding-transporter	?	at4g39670	unknown	--
sphingosine transfer protein	ACD11	At2g34690	Cytosol	Brodersen et al. 2002

Table 1. (Continued)

ENZYMES	Name	Gene	Location	References
ketosphinganine reductase	KSR	At3g06060	ER	Chao et al. 2011
		At5g19200		Chao et al. 2011
LCB C4-hydroxylase	SBH1	At1g69640	ER/Golgi	Sperling et al. 2001
	SBH2	At1g14290		Chen et al. 2008
LCB Δ8-desaturase	SLD1, DES8	At3g61580	ER/Golgi	Sperling et al. 1998
		At2g46210		García-Maroto et al. 2007
				Marion et al. 2008
LCB Δ4-desaturase	DES4	At4g04930	unknown	Ternes et al. 2002
				Bedia et al. 2005
ceramide kinase	ACD5	At5g51290	unknown	Liang et al. 2003
sphingosine-1P phosphatase	SSPase	At3g58490	unknown	Worrall et al. 2008
				Nakamura et al. 2007
sphingosine-1P lyase	DPL1	At1g27980	ER	Tsegaye et al. 2007
	SPHK1	At4g21540	Tonoplast	Worrall et al. 2008
sphingosine kinase	SPHK2	At2g46090	Mitochondria	Imai and Nishiura 2005
LCB kinase	LCBK1	At5g23450	Plasma membrane	Benschop 2007
	LCBK2	At5g51290	unknown	

Subcellular purification of different intracellular membranes has revealed different lipid composition (Moreau, Bessoule et al. 1998; Warnecke and Heinz 2003). In particular, phospholipids are the most abundant species in membranes, whereas sphingolipids and sterols can be very low or absent (such as in the ER membranes), or representing up to 25% each of total lipid levels (as in the plasma membrane). Apart from the plasma membrane, sphingolipids are enriched also in the late compartments and in the plant tonoplast, where they represent 10-20% of the total membrane lipids (Figure 4) (Moreau, Bessoule et al. 1998; Warnecke and Heinz 2003). Golgi and late secretory compartments contain discrete amounts of both sterols and sphingolipids, with a composition showing more similarity to the PM than to the ER membranes. Membrane lipids appear thus compartmentalized in the cell with a specific gradient of sphingolipid and sterol along the secretory pathway with the highest accumulation at the PM. Interestingly, while sphingolipids are enriched in the late compartments of the secretory pathway, most of the biosynthesis is localized to the ER (Table 1). There are however few exceptions with the glycosylation of Ceramide to form GluCer that occurs in the cytosolic leaflet of early Golgi cisternae. Once formed, GluCer is the transported across the Golgi either via a vesicular pathway or with the lipid transfer protein known as FAPP2 (D'Angelo, Uemura et al. 2013). Interestingly, these two differently transported GluCer pools follow two different glycosylation mechanisms, and in different Golgi subdomains. In particular, GluCer transported by direct protein transfer is used to synthesize glycosphingolipids known as globosides in the TGN while vesicular transported GluCer is used in the Golgi cisternae for ganglioside production, such as that of GM3 (D'Angelo, Uemura et al. 2013).

Table 2. Ceramide synthases

	Enzyme	Gene	FA substrate
Human	CerS1	Lass1	C18
	CerS2	Lass2	C22-C24
	CerS3	Lass3	C18,C24, all medium chain
	CerS4	Lass4	C20, few C22-C24
	CerS5	Lass5	C16
	CerS6	Lass6	C14-C16
S. Cerevisiae	Lag1	LAG1	Mostly C26
	Lac1	LAC1	Mostly C26
A. Thaliana	LOH1	loh1	C22-C26
	LOH2	loh2	Mostly C16
	LOH3	loh3	C22-C26

5. Sphingolipid Homeostasis in Cell Growth

Pools of different LCBs, ceramide and complex sphingolipids are highly dynamic. In mammals, activation of sphingomyelinase and ceramidase resulted in the accumulation of ceramide and LCBs respectively, both involved in growth arrest and cell death (Mao and Obeid 2008). Several studies indicateed that high levels of LCBs (especially sphingosine, SPH) and ceramides were associated with growth arrest, while phosphorylated sphingosine (S1P) induced cell proliferation (Ping and Barrett 1998). Thus, the relative level of ceramide and SPH versus S1P seems to associate with opposite cellular effects leading to the idea that sphingolipids may act as a rheostat system regulating the cellular switch between survival or death, reviewed by (Hannun and Obeid 2008).

Catabolism of complex sphingolipids is better understood in humans, and several autosomal recessive mutations have been reported and classified as sphingolipidosesbecause they lead to sphingolipid accumulation. Degradation of ceramide is achieved by ceramidase, sphingomylein by sphingomyelinase, glucosylceramide by glucocerebrosidase, and galactosylceramide by galactosylceramidases. Mutations of glucosylceramidase induce the Gaucher disease, which is the most common lysosomal storage disease. In patients with Gaucher disease, GluCer accumulates in spleen, liver, kidneys, lungs and brain, eventually causing organ degeneration (Grabowski, Gatt et al. 1990). Genetic defects of sphingomyelinase causes Niemann-Pick type A and B diseases, characterized by the accumulation of sphingomyelin and hepatic and neuronal degeneration (Schuchman 2010). Ceramidase mutations cause the Faber disease, characterized by progressive body deformation and, 20 pain because of ceramide accumulation in joints and the development of granulomas (Ozkara; Molino and Galli 2014).

In plants the degradative enzymes for GluCer and GIPC (analog of sphingomyelinase) have not yet been found. Three putative ceramidases (CDase) are present in the Arabidopsis genome (Table 1), but they have not been characterized yet. The only plant ceramidase so far characterized is the rice OsCDase1. It localizes to the ER and Golgi and can hydrolyze a fluorescent ceramide analog *in vitro* (Pata, Wu et al. 2008). However, no ceramidase activity against endogenous Cer was detected. On the contrary, OsCDase1 expression in yeast CDases knockout induced an increase of C-26-ceramide. This observation suggested that OsCDase1, displays a reverse ceramidase activity, resulting in an Acyl-CoA independent ceramide synthesis, similar observation where made for yeast YPC1p and YDC1p (Jiang, Kirchman et al. 2004).

The balance between ceramide synthesis and degradation appears to be essential for the homeostasis of the three signaling molecules Cer, SPH and S1P. Several stresses induce degradation of sphingomyelin and accumulation of ceramide. Known factor that have been associated with ceramede accumumation are: proapoptotic cytokine (TNF-α, IL-1β, FAS ligand, interferon γ), chemotherapeutic agents that induces DNA damage-mediated apoptosis, UV, ionizing irradiation and serum deprivation (Mao and Obeid 2008). Sphingosine and ceramide have been reported to induce both cell growth arrest and apoptosis (Pettus, Bielawski et al. 2003), (Futerman and Hannun 2004). In particular, in several carcinogenic cell lines, SPH was found to be an efficient inducer of cell death. SPH inhibits protein kinase C (PKC), the tyrosine kinase SRC (sarcoma), several Ca^{++}-Calmodulin dependent kinases (Jefferson and Schulman 1988; Igarashi, Hakomori et al. 1989) and mitogen activated kinases

ERK1/2 (Jarvis, Fornari et al. 1997). Apart from targeting cytosolic proteins, SPH also alters membrane systems. High levels of SPH increase permeability of lysosomes (Kagedal, Zhao et al. 2001) and mitochondria (Cuvillier and Levade 2001) which respectively release cathepsins and Cytochrome C to the cytosol that in turn induces apoptosis. Treatment of Hela cells with C-6 ceramide induces endogenous increase of SPH followed by Golgi fragmentation and apoptosis (Hu, Xu et al. 2005). Overexpression of ceramidases also caused SPH increases and a similar effect on Golgi fragmentation leading to apoptosis (Markham, Li et al. 2006).

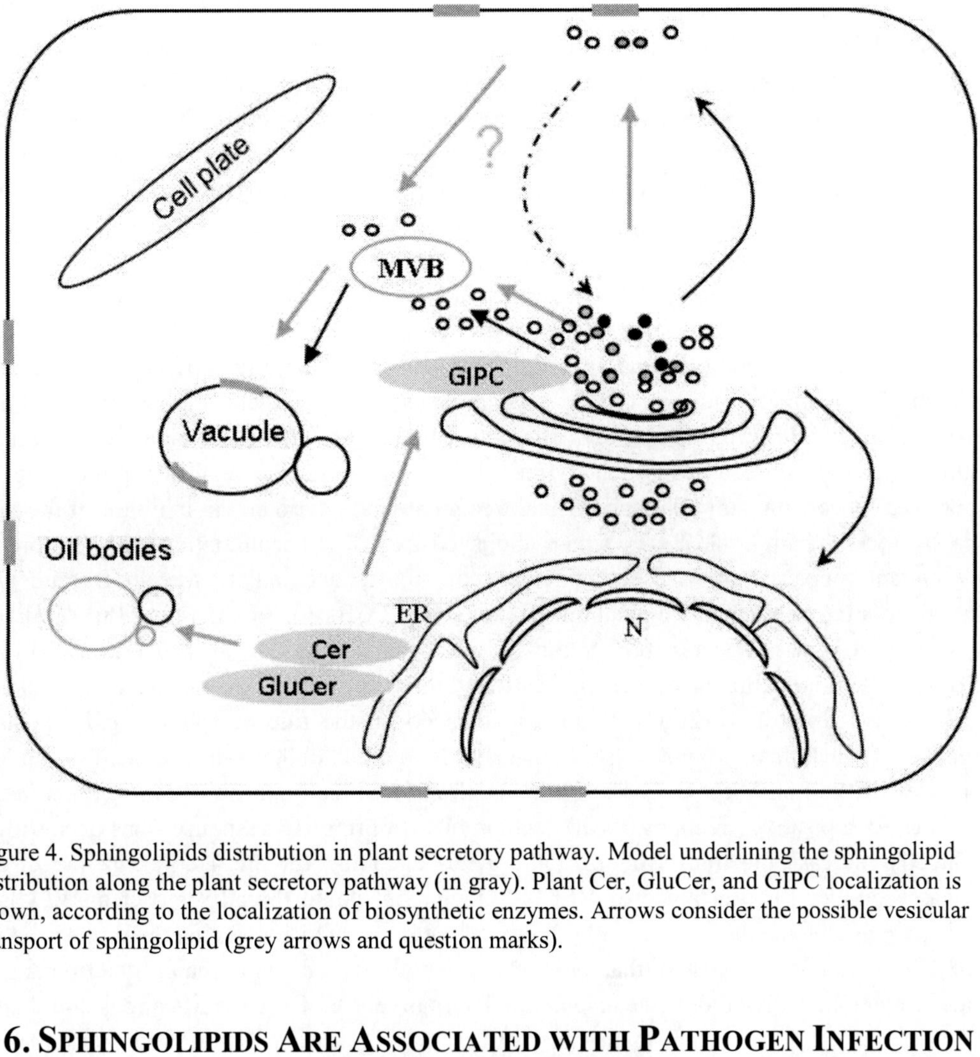

Figure 4. Sphingolipids distribution in plant secretory pathway. Model underlining the sphingolipid distribution along the plant secretory pathway (in gray). Plant Cer, GluCer, and GIPC localization is shown, according to the localization of biosynthetic enzymes. Arrows consider the possible vesicular transport of sphingolipid (grey arrows and question marks).

6. Sphingolipids Are Associated with Pathogen Infection in Plants and Animals

Sphingolipids play roles in signal transduction pathways regulating stress responses in eukaryotes. In plants LCB modifications including phosphorylation as well as unsaturation were shown to take part in the tolerance to environmental extreme and responses to chilling (Chen, Markham et al. 2012; Dutilleul, Benhassaine-Kesri et al. 2012). In the other side, modulation of the sphingolipid metabolism was associated with responses to heat stress in yeast (Sun, Miao et al. 2012).

Several Arabidopsis mutations modifying sphingolipid homeostasis enhanced sensitivity to pathogen infections. Upon infection, plant responses include generation of H_2O_2 at fungal penetration sites and triggering of the hypersensitive response (HR). HR is a form of programmed cell death at the infection site (Levine, Tenhaken et al. 1994). The gene *RPW8* confers resistance to powdery mildew, a disease caused by several fungi targeting hundreds of dicotyledonous plants. In Arabidopsis, the Columbia (Col 0) accession does not have a *RPW8* gene while WS accession contains two *RPW8* copies. EMS mutagenesis on a Col 0 background overexpressing RPW8 enabled to find mutations that suppress the resistance conferred by RPW8 (Wang, Yang et al. 2008). Such a screen identified the plant IPC-synthase ERH1, as already mentioned, *erh1* mutation prompted the accumulation of high level of ceramides explaining the spontaneous induction of cell death. Similarly, the ceramide kinase ACD5 (for accelerated cell death) (Brodersen, Petersen et al. 2002), as well as a sphingosine transfer protein ACD11 (Liang, Yao et al. 2003), were identified in a screen for mutations that cause HR responses in the absence of pathogens. Altogether these observations indicated that any mutation resulting in increased levels of ceramide, possibly at the ER, induces cell growth arrest and PCD in plant as in mammals. Ceramide phosphorylation, or its transport out of the ER, seems to be protective factors against HR, since they would prevent Cer accumulation (Liang, Yao et al. 2003).

As already mentioned, fungal infection was shown to directly modify Cer levels by inhibition of ceramide synthase activity. Mycotoxins such as AAL and FB1 inhibit CerS and induce cell death. Several pieces of evidence suggested that in plants, the bioactive molecules leading to cell death are accumulated free LCBs. AAL induced death symptoms can be suppressed by concomitant pharmacological treatment with Myriocin, an inhibitor of the SPT (Spassieva, Markham et al. 2002), which abolished free LCB accumulation. Furthermore an *lcb1* mutant, impaired in SPT activity, was not able to accumulate free LCBs and was resistant to FB1 treatment. Application of exogenous LCBs (d18:0, t:18:0, t:18:1) induced ROS (reactive oxygen species) production and pro-death responses (Shi and Wei 2007). The LCB-dependent induction of cell death was found to be dependent upon nuclear Ca^{++} release (Lachaud, Da Silva et al. 2009). Studies in tobacco on the role of sphingolipids in plant immunity against mycotoxins suggest that LCBs modulate the salicylic acid-dependent responses (Rivas-San Vicente, Larios-Zarate et al. 2013). Finally, a MYB transcription factor (AtMYB30), a positive regulator of cell death in plants during HR responses, has been shown to directly increase the transcription of the genes encoding for the 4 core enzyme of the elongase complex. The increase of VLCFA elongation would directly impact on VLCFA-bearing lipids among them the sphingolipids, as demonstrated by (Raffaele, Vailleau et al. 2008). The authors suggested that acyl chain length would be a major determinant in modulating cell death and defense responses. Alternatively VLCFA modifications could alter pathogen recognition at the membrane by modifying structural properties of plasma membrane lipids. It can also not be excluded that the increase of VLCFA simply enhanced accumulation of epicuticular waxes and increase the epidermal physical barriers preventing pathogen entry.

Interestingly enough, human viruses are known to use membranes and their components (including sphingolipids) in different steps of their life cycle including attachment, intracellular transport, and budding, reviewed in(Schneider-Schaulies and Schneider-Schaulies 2015). Examples for sphingolipid-dependent virus entry are found for human immunodeficiency virus (HIV), which besides its protein receptors also interacts with glycosphingolipids (GSLs) (Sorice, Garofalo et al. 2001) as well as the rhinovirus, which promotes the formation of ceramide- and glycosphingolipids-enriched platforms (Dreschers, Franz et al. 2007). Furthermore viruses can also manipulate sphingolipid metabolism as for example influenza A virus, which activates sphingosine kinase 1 and the transcription factor NF-κB (Seo, Pritzl et al. 2013).

7. Sphingolipid Phosphorylation: The Sphingosine-P Paradigm in Animals and Plants

Phosphorylated LCBs (LCB-P) and corresponding kinases have been so far identified in all organisms. In yeast reversible phosphorylation is carried out by two LCB-kinases, LCB4p and LCB5p (Nagiec, Skrzypek et al. 1998) while dephosphorylation is carried out by the LCB phosphatases LCB3p and YSR3p (Figure 5). LCB-P can also be degraded by the LCB phosphate-lyase DPL11p, which irreversibly breaks down the LCBs-P in ethanolamine-phosphate and a hexadecanal aldhyde (Figure 5). While LCB3p, YSR3p and DPL1p are localized to perinuclear and cortical ER, LCB4p and LCB5p are found only at the cortical ER and Golgi (Hait, Fujita et al. 2002; Iwaki, Sano et al. 2007). Single mutant strains for any of these LCB-P related genes resulted in wild type phenotype. On the contrary double mutant *dpl1/lcb3* was not viable but could be rescued by *lcb4* mutation in the triple *dpl1/lcb3/lcb4* mutant (Nagiec, Skrzypek et al. 1998). LCB-P accumulation was thus correlated with growth inhibition (Figure 5), which was confirmed by LCB4 or LCB5 overexpression in *dpl1lcb3/lcb4* mutant that increased LCB-P levels and arrested growth (Nagiec, Skrzypek et al. 1998). A number of studies suggested that LCB-P could also have a role in several stress responses in yeast. For example heat stress induced LCB-P increase and *lcb3* mutants, who accumulate LCB-P, survived better upon heat shock, while *lcb3* overexpression resulted in lower LCB-P and higher sensitivity to temperature (Mandala, Thornton et al. 1998). Altogether these observations suggested that LCB-P plays a dual role according to its concentration in the cell. A moderated increase of LCB-P would correlate with a protective role during stress responses, while greater accumulation of LCB-P would correlate with growth arrest.

It has been reported that G1 cell cycle arrest induced by rapamycin involved lcb4p degradation (Iwaki, Sano et al. 2007). Thus, it is possible that LCB-Ps are also involved in the maintenance of G1 arrest. A similar function was proposed for the mammalian S1P. In mammals, overexpression of SPHK1, a homologue of Lcb4p, results in a decreased number of cells in G0/G1 and an increased number of cells in S phase, suggesting that LCB-P could be involved in G1-to-S transition (Jenkins and Hannun 2001).

In Arabidopsis, four LCBs kinases have been identified: Sphk1 and sphk2 (for sphingosine kinase 1 and 2), and LCBK1, LCBK2 (for LCB kinase 1 and 2) (Table 1). They have different subcellular localizations involving tonoplast, mitochondria and PM (Table 1). LCBs-P can be dephosphorylated by the phosphatases SPPase (sphingosine phosphatase) and hydrolyzed by the unique lyase atDPL1 in the ER (Tsegaye, Richardson et al. 2007). Arabidopsis *dpl1* mutant does not accumulate much LCB-P or LCB. This observation suggests that LCB-P degradation via AtDPL1 does not represent a major pathway to control LCB/LCB-P levels in normal growth conditions. However, LCB and LCB-P accumulation are associated with stress conditions and cell death. The Arabidopsis *dpl1* mutant showed a hypersensitive response to FB1, which correlates with an accumulation of both LCB and LCB-P. Thus, AtDPL1 activity seems to become important in situations of stress leading to high LCB accumulation. It has to be noted that phosphorylated LCBs cannot be directly incorporated into ceramide, but can be reversibly dephosphorylated into LCB and then incorporated into sphingolipids, providing a different pathway to control LCB-P accumulation.

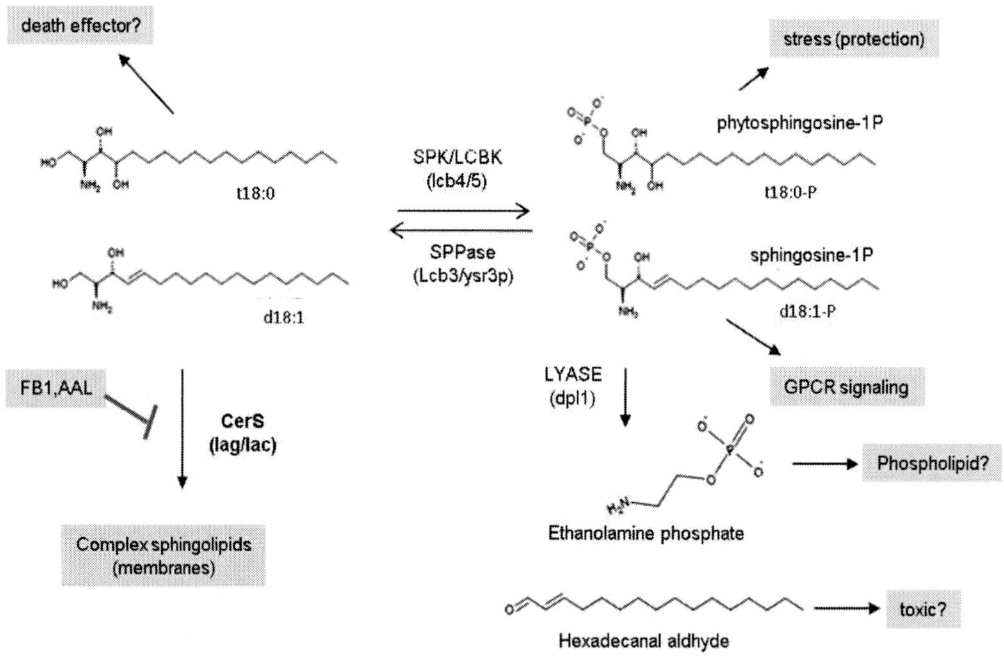

Figure 5. LCB phosphorylation pathway. LCBs can be phosphorylated and dephosphorylated by sphingosine- or LCB-kinases and sphingosine-phosphatase (SPPase) respectively. Only phosphorylated LCBs are substrate of degradation by the lyase DPL1. The interconnection of the phosphorylation pathway with the ceramide synthase is shown. In boxes, molecule functions in different organisms have been indicated. Interrogation points represent speculations.

Sphingosine phosphate (S1P) has been associated with cell proliferation in animals. Originally, S1P was shown to induce mitosis in fibroblasts (Zhang, Desai et al. 1991). Later, the pro-mitotic activity was associated with increases in Ca^{++} release *via* a non-identified mechanism which is phosphatidyl-inostol-1,4,5-P-independent (Mattie, Brooker et al. 1994). Numerous studies then revealed that S1P could directly interact at the plasma membrane with several G protein coupled receptors (GPCR), leading to their activations. At least five different S1P-GPCR mechanism mediating proliferative responses have been described in animals (Mao and Obeid 2008; Rosen, Gonzalez-Cabrera et al. 2009) that involve angiogenesis, vascular tissue development, immune responses and neuronal development. In human cell lines, S1P was also found to bind to and modulate the activity of a histone deacetylase in the nucleus (Hait, Allegood et al. 2009) and a E3 Ubiquitin ligase complex of tumor necrosis factor receptor associated factor 2 (TRAF2) in the cytosol (Alvarez, Harikumar et al. 2010). S1P signaling system was also associated with several physiological process as well as human pathology especially inflammation process and inflammatory disease, immunity cells physiology (Garris, Blaho et al. 2014; Gonzalez-Cabrera, Brown et al. 2014) as well as cancer (Zhang, Wang et al. 2014).

In plants, the pharmacological treatment with S1P induces stomata closure (Ng, Carr et al. 2001). The S1P effect was abolished in G protein mutant, indicating that S1P activity is dependent on G protein-coupled receptor activity in plants (Coursol, Fan et al. 2003). The role of S1P was confirmed by the fact that the *sphk1* mutant or SPHK1 overexpression showed respectively lower and higher sensitivity to stomatal opening after abscissic acid stimulation (Worrall, Liang et al. 2008). The activity of the SPHK1 is correlated with stomatal sensitivity to ABA. ABA inhibition of seed germination was also much less pronounced in *sphk1* mutants. However the analysis of Arabidopsis Δ4-desaturase mutant lacking both sphingosine and S1P revealed no developmental defect (including stomata closure) (Michaelson, Zauner et al. 2009). The latest observation strongly suggests that another mechanism may be responsible for stomata closure (Lynch, Chen et al. 2009). Since SPHK1 seems to use a variety of LCBs as substrate, different free LCBs than S1P could be responsible for ABA-related responses.

8. Regulation of Sphingolipid Metabolism

The homeostasis of the different sphingolipids needs to be tightly controlled to ensure cell viability. While in plants the regulation of key enzymes of the pathway remains largely unknown, in animals and yeast several important regulatory mechanisms have been uncovered.

An important breakthrough in the understanding of homeostasis and regulation of sphingolipid metabolism was recently achieved thanks to a global yeast genetic interaction study aiming to understand the role of ORM proteins, which are involved in several human inflammatory diseases. There are 2 *ORM* genes (*ORM1* and 2) in Arabidopsis, yeast and three in humans (Han, Lone et al. 2009; Breslow 2013; Kimberlin, Majumder et al. 2013). They are ER-localized proteins involved in protein folding and unfolding responses (Han, Lone et al. 2010). ORM proteins were found to be negative regulators of the serine palmitoyl transferase (SPT), the first step of sphingolipid biosynthesis (Breslow, Collins et al. 2010). Characterization of ORM-SPT complex allowed the isolation of SAC1, a phosphoinositide phosphatase involved in protein secretion in response to nutrients in both animals and yeast (Rohde and Cardenas 2003; Blagoveshchenskaya and Mayinger 2009). Interestingly, while both SAC1 and ORMs have a negative effect on SPT, they bind SPT independently, suggesting that they probably act as two independent regulatory mechanisms. Conversely, yeast *orm1/2* presented an increased SPT activity and an accumulation of LCB but not ceramides. This observation suggests that LCB and ceramide pools are not simply interconnected and in balance, but that CerS activity is also tightly regulated (Breslow, Collins et al. 2010). In Arabidopsis, small subunits of SPT (ssSPT) also limit sphingolipid biosynthesis and therefore modulate FB1 sensitivity and are essential for male gametophyte viability (Kimberlin, Majumder et al. 2013).

The activity of ORM and its inhibitory effect on SPT depends upon ORM hyperphosphorylation. Interestingly, ORM hyperphosphorylation is also sensitive to ceramide synthesis. In yeast ceramide synthase mutant *lac1*, ORMs phosphorylation was increased and SPT activity decreased (Breslow, Collins et al. 2010). This important observation suggests that ceramide levels could regulate ORM activity, defining a negative feedback loop that prevents ceramide accumulation.

In yeast, ceramide synthase activity requires another cofactor, LIP1 (for lag1/lac1 interacting protein), essential for CerS activity. LIP1 copurified with Lag1p and Lac1p (Vallee and Riezman 2005). *In vivo* labeling experiments showed that *lip1* was impaired in ceramide synthesis, also with a sphingolipid profile similar to *lac1/lag1*. The exact mechanism by which LIP1 works is not clear but it has been suggested that LIP1 could be important for specifying ceramide synthase substrate specificities. In fact, the purification of ceramide synthase complex revealed several protein associations: two Lip1p proteins could bind Lag1p–Lac1p heterodimers or Lag1p/Lac1p homodimers (Vallee and Riezman 2005). Thus, the different complexes could have different enzymatic activities, i.e., different specificities for acyl chain length or LCB.

Plant and animal genomes do not seem to encode any LIP1p related genes (data not shown). However, a different mechanism involving enzymatic channeling could regulate sphingolipid biosynthesis. It was recently demonstrated that the VLCFA-ceramide synthase CerS2, but not the LCFA-CerS4, interacted specifically with elongase enzymes allowing efficient incorporation of very long acyl chain-CoA into sphingolipids (Ohno, Suto et al. 2010). In yeast, the target of rapamycin complex 2 (TORC2) was recently found to control *de novo* ceramide synthesis (Aronova, Wedaman et al. 2008). Arnova et al. isolated a conditional allele with a mutation in AVO3 (*avo3*), a subunit of the TORC2 complex that grows normally at 25°C, but slows down at 30°C. The authors found that a few hours after transfer to 30° the synthesis of ceramides was strongly reduced. Furthermore the overexpression of a target of TOR, the kinase YPK2, reverted *avo3* phenotypes and ceramide depletion, suggesting that YPK2 kinase activity is acting in the pathway downstream of TOR and upstream of ceramide synthase. Another interesting result regarded calcineurin, a conserved Ca++/Calmodulin-regulated phosphatase that inhibits several targets of TOR. Deletion of regulatory subunit of the calcineurin could also restore proper ceramide synthesis in *avo3* (Aronova, Wedaman et al. 2008).

In conclusion, TOR is a kinase involved in stress sensing in all eukaryotes and a master regulator acting upstream Orms proteins in the signaling cascade regulating sphingolipids metabolism (Liu, Huang et al. 2012). Such a master regulator also controls actin polarization and endocytosis via unclarified mechanisms (Rispal, Eltschinger et al. 2015). Sphingolipids were also shown to be required for endocytosis and actin organization (Zanolari, Friant et al. 2000).

Interestingly enough, processes that maintain lipid and protein clustering and thus the formation of membrane domains have been proposed to be involved in cortical actin (Goswami, Gowrishankar et al. 2008). Thus, it is plausible that the control of membrane lipid biosynthesis is a basic process regulating membrane composition, tension and organization necessary for different plasma membrane activities, as well as intracellular trafficking pathways.

REFERENCES

Alvarez, S. E., Harikumar K. B., et al. (2010). Sphingosine-1-phosphate is a missing cofactor for the E3 ubiquitin ligase TRAF2. *Nature* 465(7301): 1084-8.

Aronova, S., Wedaman K., et al. (2008). Regulation of ceramide biosynthesis by TOR complex 2. *Cell Metab* 7(2): 148-58.

Bach, L., Michaelson L., et al. (2008). The plant very long chain hydroxy fatty Acyl-CoA dehydratase PASTICCINO2 is essential and limiting for plant development. *Proc. Natl. Acad. USA* 105(38): 14727-14731.

Barz, W. P. and Walter P. (1999). Two endoplasmic reticulum (ER) membrane proteins that facilitate ER-to-Golgi transport of glycosylphosphatidylinosi- tol-anchored proteins. *Mol Biol Cell* 10(4): 1043-59.

Beaudoin, F., Wu X., et al. (2009). Functional characterization of the Arabidopsis beta-ketoacyl-coenzyme A reductase candidates of the fatty acid elongase. *Plant Physiol* 150 (3): 1174-91.

Bedia, C., Triola G., et al. (2005). Analogs of the dihydroceramide desaturase inhibitor GT11 modified at the amide function: Synthesis and biological activities. *Org Biomol Chem* 3(20): 3707-12.

Benschop, J. J., Mohammed S., et al. (2007). Quantitative phosphoproteomics of early elicitor signaling in Arabidopsis. *Mol Cell Proteomics* 6(7): 1198-214.

Blagoveshchenskaya, A. and Mayinger P. (2009). SAC1 lipid phosphatase and growth control of the secretory pathway. *Mol Biosyst* 5(1): 36-42.

Breslow, D. K. (2013). Sphingolipid homeostasis in the endoplasmic reticulum and beyond. *Cold Spring Harb Perspect Biol* 5(4): a013326.

Breslow, D. K.,. Collins S. R, et al. (2010). Orm family proteins mediate sphingolipid homeostasis. *Nature* 463(7284): 1048-53.

Brodersen, P., Petersen M., et al. (2002). Knockout of Arabidopsis accelerated-cell-death11 encoding a sphingosine transfer protein causes activation of programmed cell death and defense. *Genes Dev* 16(4): 490-502.

Chao, D. Y., Gable K., et al. (2011). Sphingolipids in the root play an important role in regulating the leaf ionome in Arabidopsis thaliana. *Plant Cell* 23(3): 1061-81.

Chen, M., Han G., et al. (2006). The Essential Nature of Sphingolipids in Plants as Revealed by the Functional Identification and Characterization of the Arabidopsis LCB1 Subunit of Serine Palmitoyltransferase. *Plant Cell* 18(12): 3576-93.

Chen, M., Markham J. E., et al. (2012). Sphingolipid Delta8 unsaturation is important for glucosylceramide biosynthesis and low-temperature performance in Arabidopsis. *Plant J* 69(5): 769-81.

Chen, M., Markham J. E., et al. (2008). Sphingolipid long-chain base hydroxylation is important for growth and regulation of sphingolipid content and composition in Arabidopsis. *Plant Cell* 20(7): 1862-78.

Coderch, L., Lopez O., et al. (2003). Ceramides and skin function. *Am J Clin Dermatol* 4(2): 107-29.

Coursol, S., Fan L. M., et al. (2003). Sphingolipid signalling in Arabidopsis guard cells involves heterotrimeric G proteins. *Nature* 423(6940): 651-4.

Cuvillier, O. and Levade T. (2001). Sphingosine 1-phosphate antagonizes apoptosis of human leukemia cells by inhibiting release of cytochrome c and Smac/DIABLO from mitochondria. *Blood* 98(9): 2828-36.

D'Angelo, G., Uemura T., et al. (2013). Vesicular and non-vesicular transport feed distinct glycosylation pathways in the Golgi. *Nature*.

D'Mello N, P., Childress A. M., et al. (1994). Cloning and characterization of LAG1, a longevity-assurance gene in yeast. *J Biol Chem* 269(22): 15451-9.

Dawkins, J. L., Hulme D. J., et al. (2001). Mutations in SPTLC1, encoding serine palmitoyltransferase, long chain base subunit-1, cause hereditary sensory neuropathy type I. *Nat Genet* 27(3): 309-12.

Dreschers, S., Franz P., et al. (2007). Infections with human rhinovirus induce the formation of distinct functional membrane domains. *Cell Physiol Biochem* 20(1-4): 241-54.

Dutilleul, C., Benhassaine-Kesri G., et al. (2012). Phytosphingosine-phosphate is a signal for AtMPK6 activation and Arabidopsis response to chilling. *New Phytol* 194(1): 181-91.

Egilmez, N. K., Chen J. B., et al. (1989). Specific alterations in transcript prevalence during the yeast life span. *J Biol Chem* 264(24): 14312-7.

Endo, K., Akiyama T., et al. (1996). Degenerative spermatocyte, a novel gene encoding a transmembrane protein required for the initiation of meiosis in Drosophila spermatogenesis. *Mol Gen Genet* 253(1-2): 157-65.

Eto, M., Bennouna J., et al. (2003). C16 ceramide accumulates following androgen ablation in LNCaP prostate cancer cells. *Prostate* 57(1): 66-79.

Futerman, A. H. and Hannun Y. A. (2004). The complex life of simple sphingolipids. *EMBO Rep* 5(8): 777-82.

Gaigg, B., Neergaard T. B., et al. (2001). Depletion of acyl-coenzyme A-binding protein affects sphingolipid synthesis and causes vesicle accumulation and membrane defects in Saccharomyces cerevisiae. *Mol Biol Cell* 12(4): 1147-60.

Garcia-Maroto, F., Garrido-Cardenas J. A., et al. (2007). Cloning and molecular characterisation of a Delta8-sphingolipid-desaturase from Nicotiana tabacum closely related to Delta6-acyl-desaturases. *Plant Mol Biol* 64(3): 241-50.

Garris, C. S., Blaho V. A., et al. (2014). Sphingosine-1-phosphate receptor 1 signalling in T cells: trafficking and beyond. *Immunology* 142(3): 347-53.

Gelderblom, W. C., Jaskiewicz K., et al. (1988). Fumonisins: Novel mycotoxins with cancer-promoting activity produced by Fusarium moniliforme. *Appl Environ Microbiol* 54(7): 1806-11.

Gelderblom, W. C., Kriek N. P., et al. (1991). Toxicity and carcinogenicity of the Fusarium moniliforme metabolite, fumonisin B1, in rats. *Carcinogenesis* 12(7): 1247-51.

Gelderblom, W. C., Lebepe-Mazur S., et al. (2001). Toxicological effects in rats chronically fed low dietary levels of fumonisin B(1). *Toxicology* 161(1-2): 39-51.

Gombos, I., Kiss E., et al. (2006). Cholesterol and sphingolipids as lipid organizers of the immune cells' plasma membrane: their impact on the functions of MHC molecules, effector T-lymphocytes and T-cell death. *Immunol Lett* 104(1-2): 59-69.

Gonzalez-Cabrera, P. J., Brown S., et al. (2014). S1P signaling: new therapies and opportunities. *F1000Prime Rep* 6: 109.

Goswami, D., Gowrishankar K., et al. (2008). Nanoclusters of GPI-anchored proteins are formed by cortical actin-driven activity. *Cell* 135(6): 1085-97.

Grabowski, G. A., Gatt S., et al. (1990). Acid beta-glucosidase: enzymology and molecular biology of Gaucher disease. *Crit Rev Biochem Mol Biol* 25 (6): 385-414.

Grison, M. S., Brocard L., et al. (2015). Specific Membrane Lipid Composition Is Important for Plasmodesmata Function in Arabidopsis. *Plant Cell.*

Guillas, I., Kirchman P. A., et al. (2001). C26-CoA-dependent ceramide synthesis of Saccharomyces cerevisiae is operated by Lag1p and Lac1p. *EMBO J* 20(11): 2655-65.

Hait, N. C., Allegood J., et al. (2009). Regulation of histone acetylation in the nucleus by sphingosine-1-phosphate. *Science* 325(5945): 1254-7.

Hait, N. C., Fujita K., et al. (2002). Lcb4p sphingoid base kinase localizes to the Golgi and late endosomes. *FEBS Lett* 532(1-2): 97-102.

Han, G., Gable K., et al. (2002). The Saccharomyces cerevisiae YBR159w gene encodes the 3-ketoreductase of the microsomal fatty acid elongase. *J Biol Chem* 277(38): 35440-9.

Han, S., Lone M. A., et al. (2009). Orm1 and Orm2 are conserved endoplasmic reticulum membrane proteins regulating lipid homeostasis and protein quality control. *Proc Natl Acad Sci USA* 107(13): 5851-6.

Han, S., Lone M. A., et al. (2010). Orm1 and Orm2 are conserved endoplasmic reticulum membrane proteins regulating lipid homeostasis and protein quality control. *Proc Natl Acad Sci USA* 107(13): 5851-6.

Hanada, K., K. Kumagai, et al. (2003). Molecular machinery for non-vesicular trafficking of ceramide. *Nature* 426(6968): 803-9.

Hannun, Y. A. and Obeid L. M. (2008). Principles of bioactive lipid signalling: Lessons from sphingolipids. *Nat Rev Mol Cell Biol* 9(2): 139-50.

Hartmann, D., Lucks J., et al. (2012). Long chain ceramides and very long chain ceramides have opposite effects on human breast and colon cancer cell growth. *Int J Biochem Cell Biol* 44(4): 620-8.

Hojjati, M. R., Li Z., et al. (2005). Serine palmitoyl-CoA transferase (SPT) deficiency and sphingolipid levels in mice. *Biochim Biophys Acta* 1737(1): 44-51.

Holland, P. W., Booth H. A., et al. (2007). Classification and nomenclature of all human homeobox genes. *BMC Biol* 5: 47.

Hu, W., R. Xu, et al. (2005). Golgi fragmentation is associated with ceramide-induced cellular effects. *Mol Biol Cell* 16(3): 1555-67.

Huitema, K., van den Dikkenberg J., et al. (2004). Identification of a family of animal sphingomyelin synthases. *EMBO J* 23(1): 33-44.

Igarashi, Y., Hakomori S., et al. (1989). Effect of chemically well-defined sphingosine and its N-methyl derivatives on protein kinase C and src kinase activities. *Biochemistry* 28(17): 6796-800.

Imai, H. and Nishiura H. (2005). Phosphorylation of sphingoid long-chain bases in Arabidopsis: functional characterization and expression of the first sphingoid long-chain base Kinase gene in plants. *Plant Cell Physiol* 46(2): 375-80.

Islam, M. N., Jacquemot M. P., et al. (2012). Sphingosine in plants: More riddles from the Sphinx? *New Phytol* 193(1): 51-7.

Iwaki, S., Sano T., et al. (2007). Intracellular trafficking pathway of yeast long-chain base kinase Lcb4, from its synthesis to its degradation. *J Biol Chem* 282(39): 28485-92.

Jarvis, W. D., Fornari F. A., Jr., et al. (1997). Coordinate regulation of stress- and mitogen-activated protein kinases in the apoptotic actions of ceramide and sphingosine. *Mol Pharmacol* 52(6): 935-47.

Jefferson, A. B. and Schulman H. (1988). Sphingosine inhibits calmodulin-dependent enzymes. *J Biol Chem* 263(30): 15241-4.

Jenkins, G. M. and Hannun Y. A. (2001). Role for *de novo* sphingoid base biosynthesis in the heat-induced transient cell cycle arrest of Saccharomyces cerevisiae. *J Biol Chem* 276 (11): 8574-81.

Jiang, J. C., Kirchman P. A., et al. (2004). Suppressor analysis points to the subtle role of the LAG1 ceramide synthase gene in determining yeast longevity. *Exp Gerontol* 39(7): 999-1009.

Joubes, J., Raffaele S., et al. (2008). The VLCFA elongase gene family in Arabidopsis thaliana: phylogenetic analysis, 3D modelling and expression profiling. *Plant Mol Biol* 67(5): 547-66.

Kagedal, K., M. Zhao, et al. (2001). Sphingosine-induced apoptosis is dependent on lysosomal proteases. *Biochem J* 359(Pt 2): 335-43.

Kimberlin, A. N., Majumder S., et al. (2013). Arabidopsis 56-amino acid serine palmitoyl-transferase-interacting proteins stimulate sphingolipid synthesis, are essential, and affect mycotoxin sensitivity. *Plant Cell* 25 (11): 4627-39.

Koybasi, S., Senkal C. E., et al. (2004). Defects in cell growth regulation by C18:0-ceramide and longevity assurance gene 1 in human head and neck squamous cell carcinomas. *J Biol Chem* 279(43): 44311-9.

Lachaud, C., Da Silva D., et al. (2009). Nuclear calcium controls the apoptotic-like cell death induced by d-erythro-sphinganine in tobacco cells. *Cell Calcium* 47: 92-100.

Lauwers, E. and Andre B. (2006). Association of yeast transporters with detergent-resistant membranes correlates with their cell-surface location. *Traffic* 7(8): 1045-59.

Lester, R. L. and Dickson R. C. (2001). High-performance liquid chromatography analysis of molecular species of sphingolipid-related long chain bases and long chain base phosphates in Saccharomyces cerevisiae after derivatization with 6-aminoquinolyl-N-hydroxysuccinimidyl carbamate. *Anal Biochem* 298(2): 283-92.

Levine, A., Tenhaken R., et al. (1994). H_2O_2 from the oxidative burst orchestrates the plant hypersensitive disease resistance response. *Cell* 79 (4): 583-93.

Levine, T. P., Wiggins C. A., et al. (2000). Inositol phosphorylceramide synthase is located in the Golgi apparatus of Saccharomyces cerevisiae. *Mol Biol Cell* 11(7): 2267-81.

Liang, H., Yao N., et al. (2003). Ceramides modulate programmed cell death in plants. *Genes Dev* 17(21): 2636-41.

Lingwood, D. and Simons K. (2010). Lipid rafts as a membrane-organizing principle. *Science* 327(5961): 46-50.

Liu, M., Huang C., et al. (2012). Regulation of sphingolipid synthesis through Orm1 and Orm2 in yeast. *J Cell Sci* 125(Pt 10): 2428-35.

Lynch, D. V., Chen M., et al. (2009). Lipid signaling in Arabidopsis: no sphingosine? No problem! *Trends Plant Sci* 14(9): 463-6.

Lynch, D. V. and Dunn T. M. (2004). An introduction to plant sphingolipids and a review of recent advances in understanding their metabolism and function. *New Phytol.* 161: 677-702.

Mandala, S. M., Thornton R., et al. (1998). Sphingoid base 1-phosphate phosphatase: A key regulator of sphingolipid metabolism and stress response. *Proc Natl Acad Sci USA* 95(1): 150-5.

Mao, C. and Obeid L. M. (2008). Ceramidases: regulators of cellular responses mediated by ceramide, sphingosine, and sphingosine-1-phosphate. *Biochim Biophys Acta* 1781(9): 424-34.

Marasas, W. F., Kriek N. P., et al. (1984). Primary liver cancer and oesophageal basal cell hyperplasia in rats caused by Fusarium moniliforme. *Int J Cancer* 34(3): 383-7.

Marion, J., Bach L., et al. (2008). Systematic analysis of protein subcellular localization and interaction using high-throughput transient transformation of Arabidopsis seedlings. *Plant J* 56(1): 169-79.

Markham, J. E. and Jaworski J. G. (2007). Rapid measurement of sphingolipids from Arabidopsis thaliana by reversed-phase high-performance liquid chromatography coupled to electrospray ionization tandem mass spectrometry. *Rapid Commun Mass Spectrom* 21 (7): 1304-14.

Markham, J. E., Li J., et al. (2006). Plant sphingolipids: Separation and identification of major sphingolipid classes from leaves. *J Biol Chem* 281(32): 22684-94.

Markham, J. E., Molino D., et al. (2011). Sphingolipids containing very-long-chain fatty acids define a secretory pathway for specific polar plasma membrane protein targeting in Arabidopsis. *Plant Cell* 23(6): 2362-78.

Mattie, M., Brooker G., et al. (1994). Sphingosine-1-phosphate, a putative second messenger, mobilizes calcium from internal stores via an inositol trisphosphate-independent pathway. *J Biol Chem* 269(5): 3181-8.

Melser, S., Batailler B., et al. (2010). Glucosylceramide biosynthesis is involved in Golgi morphology and protein secretion in plant cells. *Traffic* 11(4): 479-90.

Melser, S., Wattelet-Boyer V., et al. (2009). Blocking ER export of the Golgi SNARE SYP31 affects plant growth. *Plant Signal Behav* 4(10): 962-4.

Merrill, A. H., Jr., van Echten G., et al. (1993). Fumonisin B1 inhibits sphingosine (sphinganine) N-acyltransferase and *de novo* sphingolipid biosynthesis in cultured neurons *in situ*. *J Biol Chem* 268(36): 27299-306.

Merrill, A. H., Jr., Wang E., et al. (1993). Fumonisins and other inhibitors of *de novo* sphingolipid biosynthesis. *Adv Lipid Res* 26: 215-34.

Michaelson, L. V., Zauner S., et al. (2009). Functional characterization of a higher plant sphingolipid Delta4-desaturase: defining the role of sphingosine and sphingosine-1-phosphate in Arabidopsis. *Plant Physiol* 149(1): 487-98.

Mina, J. G., Okada Y., et al. (2010). Functional analyses of differentially expressed isoforms of the Arabidopsis inositol phosphorylceramide synthase. *Plant Mol Biol* 73(4-5): 399-407.

Mizutani, Y., Mitsutake S., et al. (2009). Ceramide biosynthesis in keratinocyte and its role in skin function. *Biochimie* 91(6): 784-90.

Molino, D. and Galli T. (2014). Biogenesis and transport of membrane domains-potential implications in brain pathologies. *Biochimie* 96: 75-84.

Molino, D., Van der Giessen E., et al. (2014). Inhibition of very long acyl chain sphingolipid synthesis modifies membrane dynamics during plant cytokinesis. *Biochim Biophys Acta* 1842(10): 1422-30.

Moreau, P., Bessoule J. J., et al. (1998). Lipid trafficking in plant cells. *Prog Lipid Res* 37(6): 371-91.

Nagiec, M. M., Skrzypek M., et al. (1998). The LCB4 (YOR171c) and LCB5 (YLR260w) genes of Saccharomyces encode sphingoid long chain base kinases. *J Biol Chem* 273(31): 19437-42.

Nakamura, T., Yonesu K., et al. (2007). Synthesis and SAR studies of a novel class of S1P1 receptor antagonists. *Bioorg Med Chem* 15(10): 3548-64.

Ng, C. K., Carr K., et al. (2001). Drought-induced guard cell signal transduction involves sphingosine-1-phosphate. *Nature* 410(6828): 596-9.

Oh, C. S., Toke D. A., et al. (1997). ELO2 and ELO3, homologues of the Saccharomyces cerevisiae ELO1 gene, function in fatty acid elongation and are required for sphingolipid formation. *J Biol Chem* 272(28): 17376-84.

Ohno, Y., Suto S., et al. (2010). ELOVL1 production of C24 acyl-CoAs is linked to C24 sphingolipid synthesis. *Proc Natl Acad Sci USA* 107(43): 18439-44.

Osawa, Y., Uchinami H., et al. (2005). Roles for C16-ceramide and sphingosine 1-phosphate in regulating hepatocyte apoptosis in response to tumor necrosis factor-alpha. *J Biol Chem* 280(30): 27879-87.

Ozkara, H. A. (2004). Recent advances in the biochemistry and genetics of sphingolipidoses. *Brain Dev* 26(8): 497-505.

Pata, M. O., Hannun Y. A., et al. (2010). Plant sphingolipids: Decoding the enigma of the Sphinx. *New Phytol* 185(3): 611-30.

Pata, M. O., Wu B. X., et al. (2008). Molecular Cloning and Characterization of OsCDase, a Ceramidase Enzyme from Rice. *Plant J.*

Pettus, B. J., Bielawski J., et al. (2003). The sphingosine kinase 1/sphingosine-1-phosphate pathway mediates COX-2 induction and PGE2 production in response to TNF-alpha. *FASEB J* 17(11): 1411-21.

Pewzner-Jung, Y., Ben-Dor S., et al. (2006). When do Lasses (longevity assurance genes) become CerS (ceramide synthases)? Insights into the regulation of ceramide synthesis. *J Biol Chem* 281(35): 25001-5.

Pewzner-Jung, Y., Park, H et al. (2010). A critical role for ceramide synthase 2 in liver homeostasis: I. alterations in lipid metabolic pathways. *J Biol Chem* 285(14): 10902-10.

Pienaar, J. G., Kellerman T. S., et al. (1981). Field outbreaks of leukoencephalomalacia in horses consuming maize infected by Fusarium verticillioides (= F. moniliforme) in South Africa. *J S Afr Vet Assoc* 52(1): 21-4.

Ping, S. E. and Barrett G. L. (1998). Ceramide can induce cell death in sensory neurons, whereas ceramide analogues and sphingosine promote survival. *J Neurosci Res* 54(2): 206-13.

Ponnusamy, S., Meyers-Needham M., et al. (2010). Sphingolipids and cancer: ceramide and sphingosine-1-phosphate in the regulation of cell death and drug resistance. *Future Oncol* 6(10): 1603-24.

Rabie, C. J., Marasas W. F., et al. (1982). Moniliformin production and toxicity of different Fusarium species from Southern Africa. *Appl Environ Microbiol* 43(3): 517-21.

Raffaele, S., Vailleau F., et al. (2008). A MYB Transcription Factor Regulates Very-Long-Chain Fatty Acid Biosynthesis for Activation of the Hypersensitive Cell Death Response in Arabidopsis. *Plant Cell* 20(3): 752-67.

Riebeling, C., Allegood J. C., et al. (2003). Two mammalian longevity assurance gene (LAG1) family members, trh1 and trh4, regulate dihydroceramide synthesis using different fatty acyl-CoA donors. *J Biol Chem* 278(44): 43452-9.

Rispal, D., Eltschinger S., et al. (2015). Target of rapamycin complex 2 regulates actin polarization and endocytosis via multiple pathways. *J Biol Chem.*

Rivas-San Vicente, Larios-Zarate M., G., et al. (2013). Disruption of sphingolipid biosynthesis in Nicotiana benthamiana activates salicylic acid-dependent responses and compromises resistance to Alternaria alternata f. sp. lycopersici. *Planta* 237(1): 121-36.

Rohde, J. R. and Cardenas M. E. (2003). The tor pathway regulates gene expression by linking nutrient sensing to histone acetylation. *Mol Cell Biol* 23(2): 629-35.

Rosen, H., Gonzalez-Cabrera P. J., et al. (2009). Sphingosine 1-phosphate receptor signaling. *Annu Rev Biochem* 78: 743-68.

Schneider-Schaulies, J. and Schneider-Schaulies S. (2015). Sphingolipids in viral infection. *Biol Chem* 396(6-7): 585-95.

Schorling, S., Vallee B., et al. (2001). Lag1p and Lac1p are essential for the Acyl-CoA-dependent ceramide synthase reaction in Saccharomyces cerevisae. *Mol. Biol.Cell.* 12 (11): 3417-27.

Schuchman, E. H. (2010). Acid sphingomyelinase, cell membranes and human disease: lessons from Niemann-Pick disease. *FEBS Lett* 584(9): 1895-900.

Senkal, C. E., Ponnusamy S., et al. (2011). Alteration of ceramide synthase 6/C16-ceramide induces activating transcription factor 6-mediated endoplasmic reticulum (ER) stress and apoptosis via perturbation of cellular Ca^{2+} and ER/Golgi membrane network. *J Biol Chem* 286(49): 42446-58.

Seo, Y. J., Pritzl C. J., et al. (2013). Sphingosine kinase 1 serves as a pro-viral factor by regulating viral RNA synthesis and nuclear export of viral ribonucleoprotein complex upon influenza virus infection. *PLoS One* 8(8): e75005.

Shi, J. and Wei L. (2007). Rho kinase in the regulation of cell death and survival. *Arch Immunol Ther Exp (Warsz)* 55(2): 61-75.

Sorice, M., Garofalo T., et al. (2001). Evidence for cell surface association between CXCR4 and ganglioside GM3 after gp120 binding in SupT1 lymphoblastoid cells. *FEBS Lett* 506(1): 55-60.

Spassieva, S. D., Markham J. E., et al. (2002). The plant disease resistance gene Asc-1 prevents disruption of sphingolipid metabolism during AAL-toxin-induced programmed cell death. *Plant J* 32(4): 561-72.

Sperling, P., Franke S., et al. (2005). Are glucocerebrosides the predominant sphingolipids in plant plasma membranes? *Plant Physiol Biochem* 43(12): 1031-8.

Sperling, P. and Heinz E. (2003). Plant sphingolipids: structural diversity, biosynthesis, first genes and functions. *Biochim Biophys Acta* 1632(1-3): 1-15.

Sperling, P., Ternes P., et al. (2001). Functional characterization of sphingolipid C4-hydroxylase genes from Arabidopsis thaliana. *FEBS Lett* 494(1-2): 90-4.

Sperling, P., Zahringer U., et al. (1998). A sphingolipid desaturase from higher plants. Identification of a new cytochrome b5 fusion protein. *J Biol Chem* 273(44): 28590-6.

Sun, Y., Miao Y., et al. (2012). Orm protein phosphoregulation mediates transient sphingolipid biosynthesis response to heat stress via the Pkh-Ypk and Cdc55-PP2A pathways. *Mol Biol Cell* 23(12): 2388-98.

Tafesse, F. G., Huitema K., et al. (2007). Both sphingomyelin synthases SMS1 and SMS2 are required for sphingomyelin homeostasis and growth in human HeLa cells. *J Biol Chem* 282(24): 17537-47.

Tamura, K., Mitsuhashi N., et al. (2001). Characterization of an Arabidopsis cDNA encoding a subunit of serine palmitoyltransferase, the initial enzyme in sphingolipid biosynthesis. *Plant Cell Physiol* 42(11): 1274-81.

Tellier, F., Maia-Grondard A., et al. (2014). Comparative plant sphingolipidomic reveals specific lipids in seeds and oil. *Phytochemistry* 103: 50-8.

Teng, C., Dong H., et al. (2008). Serine palmitoyltransferase, a key enzyme for *de novo* synthesis of sphingolipids, is essential for male gametophyte development in Arabidopsis. *Plant Physiol* 146(3): 1322-32.

Ternes, P., S. Franke, et al. (2002). Identification and characterization of a sphingolipid delta 4-desaturase family. *J Biol Chem* 277(28): 25512-8.

Tsegaye, Y., Richardson C. G., et al. (2007). Arabidopsis mutants lacking long chain base phosphate lyase are fumonisin-sensitive and accumulate trihydroxy-18:1 long chain base phosphate. *J Biol Chem* 282(38): 28195-206.

Vallee, B. and Riezman H. (2005). Lip1p: A novel subunit of acyl-CoA ceramide synthase. *EMBO J* 24(4): 730-41.

Venkataraman, K. and Futerman A. H. (2002). Do longevity assurance genes containing Hox domains regulate cell development via ceramide synthesis? *FEBS Lett* 528(1-3): 3-4.

Wang, W., Yang X., et al. (2008). An inositolphosphorylceramide synthase is involved in regulation of plant programmed cell death associated with defense in Arabidopsis. *Plant Cell* 20(11): 3163-79.

Warnecke, D. and Heinz E. (2003). Recently discovered functions of glucosylceramides in plants and fungi. *Cell Mol Life Sci* 60(5): 919-41.

West, G., Viitanen L., et al. (2008). Identification of a glycosphingolipid transfer protein GLTP1 in Arabidopsis thaliana. *FEBS J* 275(13): 3421-37.

Worrall, D., Liang Y. K., et al. (2008). Involvement of sphingosine kinase in plant cell signalling. *Plant J* 56(1): 64-72.

Yamaji, T. and Hanada K. (2015). Sphingolipid metabolism and interorganellar transport: localization of sphingolipid enzymes and lipid transfer proteins. *Traffic* 16(2): 101-22.

Yoo, H. S., Norred W. P., et al. (1992). Fumonisin inhibition of *de novo* sphingolipid biosynthesis and cytotoxicity are correlated in LLC-PK1 cells. *Toxicol Appl Pharmacol* 114(1): 9-15.

Zanolari, B., Friant S., et al. (2000). Sphingoid base synthesis requirement for endocytosis in Saccharomyces cerevisiae. *EMBO J* 19(12): 2824-33.

Zhang, H., Desai N. N., et al. (1991). Sphingosine-1-phosphate, a novel lipid, involved in cellular proliferation. *J Cell Biol* 114(1): 155-67.

Zhang, Y., Wang Y., et al. (2014). Sphingosine kinase 1 and cancer: a systematic review and meta-analysis. *PLoS One* 9(2): e90362.

Zheng, H., Rowland O., et al. (2005). Disruptions of the Arabidopsis Enoyl-CoA reductase gene reveal an essential role for very-long-chain fatty acid synthesis in cell expansion during plant morphogenesis. *Plant Cell* 17(5): 1467-81.

In: Sphingolipids
Editor: Angel Catalá

ISBN: 978-1-63483-019-5

Chapter 4

METABOLISM OF GLYCOLIPIDS AT THE CELL SURFACE

Jose L. Daniotti*[*]*, Aldo A. Vilcaes, Macarena Rodriguez-Walker and Fernando M. Ruggiero

Centro de Investigaciones en Química Biológica de Córdoba (CIQUIBIC, UNC-CONICET), Departamento de Química Biológica, Facultad de Ciencias Químicas, Universidad Nacional de Córdoba, Córdoba, Argentina

ABSTRACT

Glycosphingolipids (GSLs) are amphipathic molecules consisting of a ceramide lipid moiety linked to a glycan chain of variable length and structure. Among these are found the gangliosides, which are mono- or multi-sialosylated GSLs ubiquitously distributed on the external leaflet of vertebrate plasma membranes. Gangliosides have been implicated in many physiological processes, including growth, apoptosis, cell adhesion and endocytosis, as well as in pathological processes due to their capacity to function as receptors for several toxins, viruses and antibodies. Their *de novo* synthesis starts at the endoplasmic reticulum and continues by a combination of glycosyltransferase activities at the Golgi complex, followed by vesicular delivery to the plasma membrane. At the cell surface, gangliosides can undergo endocytosis, and once internalized they can be sorted in a cell-type-dependent manner into endosomes for recycling to the plasma membrane or to be catabolized at the lysosomal level. However, a number of enzymes for ganglioside anabolism and catabolism have recently been shown to associate with the plasma membrane, being able to exert their enzymatic activities on substrates residing in the same surface (cis-catalytic activity) and in that of neighboring cells (trans-catalytic activity). The action of these enzymatic activities may be responsible for structural changes of the plasma membrane gangliosides, thus modifying membrane organization, and consequently, different signaling processes. In this Chapter, we review early and recent advances concerning the glycolipid metabolism at the cell surface, as well as its

[*] Address correspondence to: Jose Luis Daniotti, Centro de Investigaciones en Química Biológica de Córdoba (CIQUIBIC, UNC-CONICET), Departamento de Química Biológica, Facultad de Ciencias Químicas, Universidad Nacional de Córdoba, Córdoba, X5000HUA, Argentina. E-mail: daniotti@dqb.fcq.unc.edu.ar.

possible biological implications, which represents a new paradigm in the cellular control of glycolipid expression.

Keywords: sphingolipids, glycosphingolipids, gangliosides, glycolipids, glycosidases, glycosyltransferases, intracellular trafficking

INTRODUCTION

Glycolipids are molecules containing one or more carbohydrate residues linked to a hydrophobic lipid moiety via a β-glycosidic linkage, with those containing either a sphingoid or a ceramide as the hydrophobic lipid moiety referred to as GSLs. Among these are found the gangliosides, which are mono- or multi-sialosylated GSLs mainly located in the outer layer of the plasma membrane of essentially all vertebrate cells. These are expressed in cell-type- and developmental-specific patterns and are major components of nerve cells as they can represent more than 10% of the total lipid content, with temporal and spatial differences being observed at both the headgroup and lipid backbone levels [1, 2]. Gangliosides, originally identified as structural components of biomembranes, were later identified as key lipids implicated in a diverse range of physiological and pathological processes. For example, gangliosides can modulate cell signalling by regulating the activities of receptor-tyrosine kinases in the plasma membrane, such as the epidermal growth factor, the insulin and the nerve growth factor receptors. In addition, gangliosides are involved in cell-cell recognition and adhesion [3-6]. However, some ganglioside functions are undesirable, as they are receptors for viruses (i.e., simian virus 40, SV40), toxins [i.e., cholera; tetanus and botulinum toxins], lectins and antibodies [7-10]. In particular, neuropathologic diseases, such as Guillain-Barré and Miller-Fisher syndromes, are caused by an autoimmune response to gangliosides [11-13]. Aberrant glycosylation of glycolipids (including gangliosides), on the surface of the majority of cancer cells has also been widely described. Altered ganglioside expression not only has a considerable impact on signaling and cell-cell interactions, but is currently being investigated as a convenient immunotherapeutic target for cancer treatment [14].

Inherited defects in the degradation of gangliosides cause a group of severe diseases known as GSL storage disorders [15-18]. Unlike disorders of ganglioside degradation, much less is known regarding the molecular and pathophysiologic consequences of altered ganglioside biosynthesis. To date, only a few human mutations have been described in genes coding for glycosyltransferases ST3GAL-V, β4GALNT-I and ST3GAL-III.

In this Chapter, we summarize the cellular and molecular basis involved in the biosynthesis and catabolism of gangliosides, paying particular attention to recent findings concerning ganglioside metabolic remodeling at the plasma membrane.

GANGLIOSIDE BIOSYNTHESIS

The biosynthesis of gangliosides starts with the synthesis of ceramide at the cytoplasmic leaflet of the endoplasmic reticulum membrane [19]. In the first step, pyridoxal 5'-phosphate-dependent serine palmitoyltransferase catalyzes the condensation of palmitoyl- or stearoyl-

Coenzyme A with L-serine to give 3-ketosphinganine, which is reduced to D-erythro-sphinganine by 3-ketosphinganine reductase in an NADPH-dependent reaction. This compound is further acylated to generate different dihydroceramides by a family of ceramide synthases, which utilize a wide spectrum of fatty acyl-Coenzyme A. In the following step, dihydroceramide is unsaturated by members of the DES desaturase family at the 4,5 position to give ceramide [for review, see [18, 20]]. The *de novo* synthesized ceramide is then transported from the endoplasmic reticulum to the trans Golgi network, at least in part in a protein-dependent manner by the transport protein CERT (CERamide Transport), where it is catalytically converted to sphingomyelin through sphingomyelin synthase 1, which transfers the choline headgroup from phosphatidylcholine to ceramide [21]. It has also been suggested that CERT may also transport ceramide to the trans Golgi to be used for glucosylceramide (GlcCer) synthesis by the action of UDP-Glc: ceramide glucosyltransferase (UGCG) [22]. Most GlcCer may subsequently be transported by the four-phosphate adaptor protein 2 (FAPP2, a glycolipid-transport protein carrying a PI4P binding domain) either to the endoplasmic reticulum [22] or to distal Golgi compartments [23]. Since the synthesis of GlcCer occurs on the cytoplasmic face and that of lactosylceramide (LacCer) on the luminal site of the Golgi complex [24], GlcCer must be translocated across a membrane, which is mediated by a flippase activity. It should be mentioned that UGCG is a type III integral membrane protein, with an N-terminal uncleaved signal anchor sequence and a long cytoplasmic tail bearing the catalytic domain. In contrast, the glycosyltransferases involved in the building up of the oligosaccharide chain of higher order gangliosides starting from GlcCer are type II integral membrane proteins, with an N-terminal domain consisting of a relatively short cytoplasmic tail facing the cytoplasm, a transmembrane uncleaved signal-anchor region and a lumenally oriented C-terminal domain that bears the catalytic and the sugar nucleotide-binding sites [1].

The synthesis of LacCer occurs by the action of UDP-Gal:glucosylceramide galactosyl-transferase (β4GalT-VI), which transfers a galactose residue from UDP-Gal to GlcCer. Further carbohydrate residues, including sialic acid, are transferred in a stepwise manner to the growing glycan chains. Sialylated derivatives from LacCer are produced by the action of ST3Gal-V, ST8Sia-I and ST8Sia-I/ ST8Sia-V, which specifically catalyze the formation of the gangliosides GM3, GD3 and GT3, respectively (Figure 1). LacCer, GM3, GD3 and GT3 serve as precursors for more complex gangliosides of the 0-, a-, b- or c-series by sequential glycosylations catalyzed by β4GalNAcT-I, β3GalT-IV, ST3Gal-II and ST8Sia-V. These transferases are non-redundant and specific, and catalyze a sugar transfer to a glycosyl acceptor that differs only in the number of sialic acids linked to the inner galactosyl residue [none (0-series), one (a-series), two (b-series), or three (c-series)] (Figure 1). After synthesis at the Golgi complex, gangliosides are mainly delivered by vesicular transport to the plasma membrane, where they can undergo endocytosis [2, 8, 25-27]. Exceptionally, the ganglioside GM4 (NeuAcα2,3Gal-Ceramide), a major component of the myelin, is not derived from LacCer.

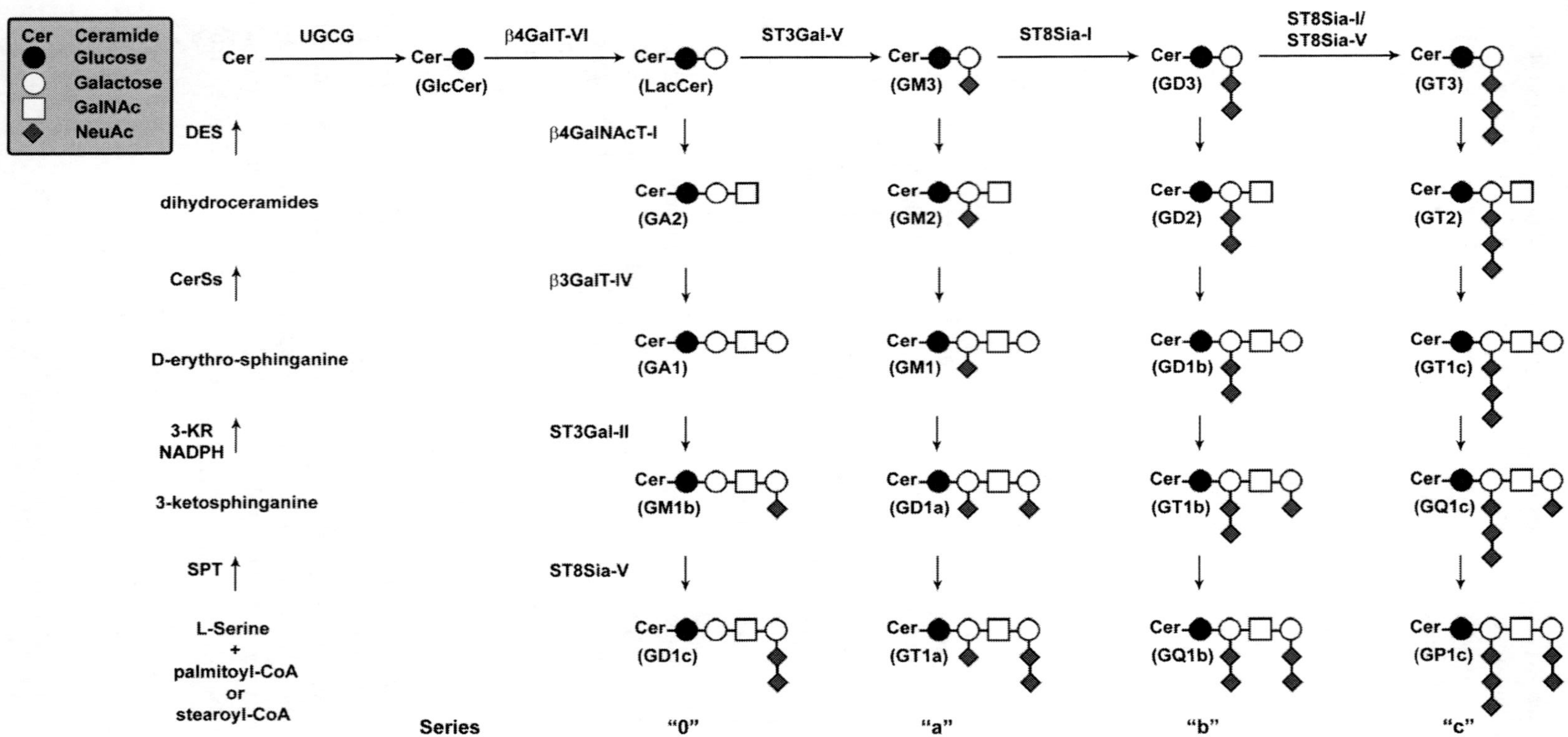

Figure 1. Biosynthesis of ceramide and gangliosides. Transfer steps for synthesis of Ceramide (Cer) and 0-, a-, b-, and c-series gangliosides are shown. SPT, pyridoxal 5'-phosphate-dependent serine palmitoyltransferase; 3-KR, 3-ketosphinganine reductase; CerSs, ceramide synthases; DES, dihydroceramide desaturase; UGCG, UDP-Glc:ceramide glucosyltransferase; β4GalT-VI, UDP-Gal:glucosylceramide galactosyltransferase; ST3Gal-V, CMP-NeuAc:lactosylceramide sialyltransferase (also called Sial-T1); ST8Sia-I, CMP-NeuAc:GM3 sialyltransferase (also called Sial-T2) and CMP-NeuAc:GD3 sialyltransferase; β4GalNAcT-I, UDP-GalNAc:lactosylceramide/GM3/GD3/GT3 N-acetylgalactosaminyl transferase (also called GalNAc-T); β3GalT-IV, UDP-Gal: GA2/GM2/GD2/GT2 galactosyltransferase (also called Gal-T2); ST3Gal-II, CMP-NeuAc:GA1/GM1/GD1b/GT1c sialyltransferase (also called Sial-T4); ST8Sia-V, CMP-NeuAc:GM1b/GD1a/GT1b/GQ1c sialyltransferase and CMP-NeuAc:GD3 sialyltransferase. Gangliosides are named according to Svennerholm [129].

In addition to the bulk ganglioside synthesis at the Golgi complex level, ganglioside formation by plasma membrane-associated glycosyltransferases has been reported. In particular, a sialyltransferase (ectosialyltransferase) and an N-acetylgalactosaminiltransferase were demonstrated at the cell surface of different cell types [28, 29]. This topic is further reviewed below.

Very little is known about deficiencies in the biosynthetic pathways of gangliosides. Human mutation in the sialyltransferase gene *ST3GAL-V* results in severe infantile-onset seizures and developmental delay [30-32]. Mutations in the brain ganglioside biosynthetic gene *β4GALNT-I* result in hereditary spastic paraplegia [33, 34], whereas mutations in the glycoprotein/ganglioside sialyltransferase gene *ST3GAL-III* result in nonsyndromic intellectual disability [35] and West syndrome [36].

Additionally, a human disease has been linked with mutations in the gene encoding serine palmitoyltransferase long chain base subunit-1, which was associated with increased *de novo* GlcCer synthesis and massive cell death during neural tube closure, consequently causing hereditary sensory neuropathy type I [37].

Ganglioside Biodegradation

Biodegradation of gangliosides take place mainly at the luminal surface of lysosomes. Additionally, degradation of gangliosides can also occur at the cell surface level by action of the sialidase Neu3, β-galactosidase and β-glucosidase, which will be further described in detail below. At the lysosomal level, gangliosides are degraded by the action of glycosidases that cleave off sequentially the monosaccharide units from the non-reducing end of the ganglioside glycan chains [15] (Figure 2). Adequate lysosomal ganglioside catabolism requires the presence of an appropriate pH, suitable glycosidases and lipid-transfer proteins for degradation of simple gangliosides. In this regard, it is assumed that GSLs with four carbohydrate residues or less require the presence of small lipid binding glycoproteins (GM2 activator protein or one of the four saposins A, B, C and D), which extract the membrane-bound glycolipids and present them to the soluble acid hydrolase [15, 17]. Additionally, the presence of anionic lipids, such as bis(monoacylglycero)phosphate, play a key role in ganglioside degradation by attracting glycosidases and lipid binding proteins to the ganglioside-containing membrane surfaces.

If a defect is inherited in the genes encoding glycosidases and lipid-transfer protein essential for the lysosomal catabolism of GSLs, the substrate is stored. In particular, degradation of gangliosides is impaired in gangliosidosis and secondarily in other sphingolipid storage diseases, with lysosomal storage diseases often being associated with severe neurodegenerative pathology. Detailed information about ganglioside catabolism and inherited sphingolipid storage disease degradation has been recently covered by some excellent reviews [17, 38-40].

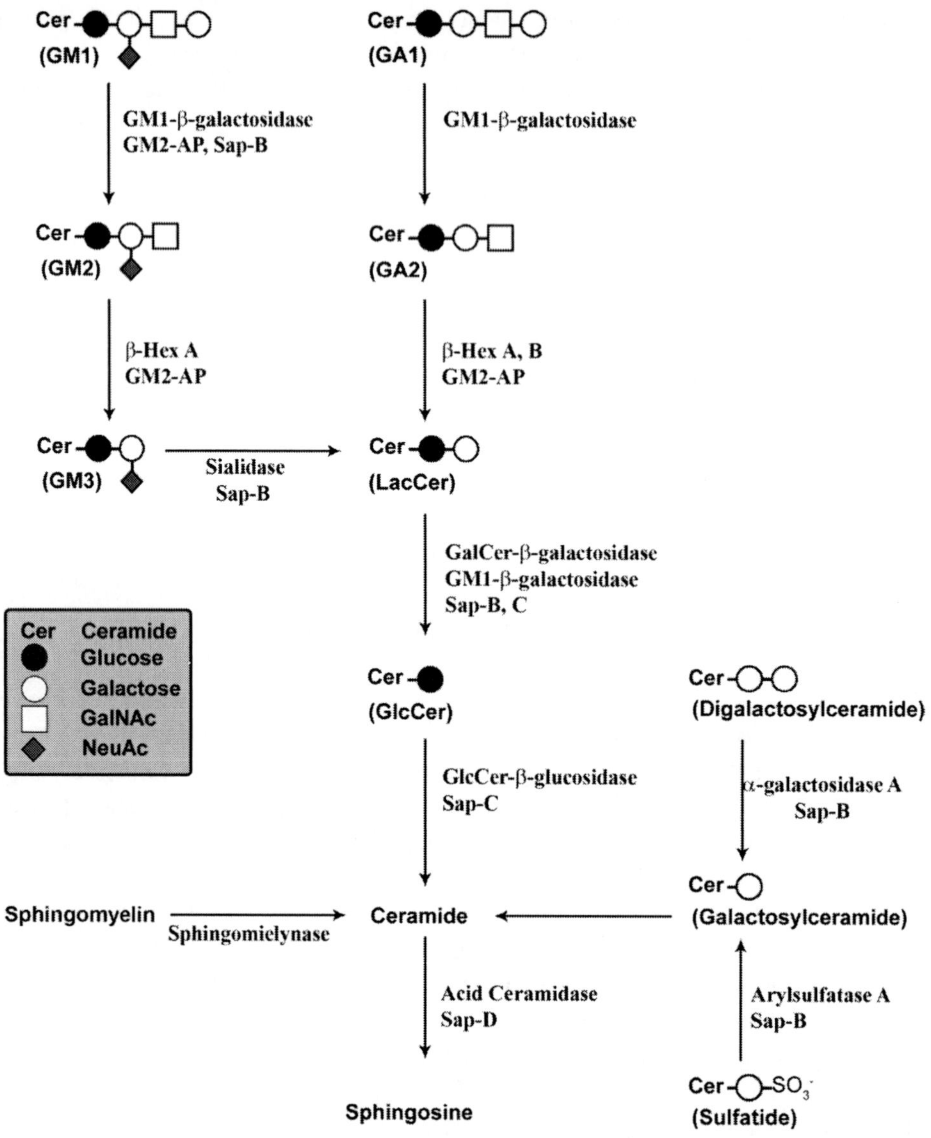

Figure 2. Biodegradation steps of gangliosides. Degradation steps from GM1 and GA1 gangliosides to sphingosine are shown. In addition the catabolism of digalactosylceramide and sulfatide is also shown. GM2-AP, GM2-activator protein; β-Hex A, β-Hexosaminidase A; β-Hex B, β-Hexosaminidase B; Sialidase (Neu1, Neu2, Neu3 and Neu4. See the text for substrate specificity); Sap-A, -B, -C and -D, Saposin-A, -B, -C and -D, respectively.

ANABOLISM OF GANGLIOSIDE AT THE CELL SURFACE

The first evidence demonstrating the presence of glycosyltransferases (specifically glycoprotein ones) at the cell surface was obtained 45 years ago [41-44]. Thus, before referring to what is known about glycolipid glycosyltranserases, we consider it opportune to briefly describe the particular glycosyltransferases responsible for the synthesis of oligosaccharides from proteins.

Cell Surface Glycoprotein Glycosyltransferases

β-1,4-Galactosyltransferase 1

β4GalT-I (UDP-Gal: β-GlcNAcβ-1,4-galactosyltransferase 1) is a type II transmembrane protein which catalyzes the incorporation of galactose from UDP-Gal to GlcNAc residues terminally located in nascent glycoproteins. In common with other glycosyltransferases, it mainly resides at the Golgi complex [45, 46], but has also been localized at the plasma membrane [47-50]. The β4GalT-I gene produces two almost identical proteins, differing only in the length of their cytoplasmic domains [11 *vs* 24 amino acids, [51]], with both being localized at the Golgi complex. However, a fraction of the long form of β4GalT-I is also targeted to the cell surface, where it functions as a cell adhesion molecule [52, 53]. This issue has attracted investigation concerning the implications on several cellular processes, such as cell-cell recognition [53, 54], cell-ECM interaction and migration [55, 56], signaling [57-59] and sperm-egg recognition [60-63]. Thus, surface β4GalT-I might be acting as a class of endogenous lectin capable of interacting in a non-catalytic mode with surface ligands.

Collagen Glucosyl- and Galactosyltransferases

Early studies reported the presence of collagen glucosyltransferase activity in plasma membrane preparation from human blood platelets [64-67]. This glycosyltransferase, which transfers glucose from UDP-glucose specifically to incomplete heterosaccharide chains of collagen, is implicated in the adhesion of platelet cells to subendothelial collagen at the beginning of hemostasis.

Collagen galactosyltransferase is another enzyme implicated in the incorporation of monosaccharide precursors into the heterosaccharide chains of collagen and has also been localized in platelet plasma membranes. However, unlike collagen glucosyltransferase, the galactosyltransferase is probably not involved in platelet: collagen adhesion [64].

Sialyltransferases

Ectosialyltransferase activity has been observed in human B cells [68, 69], with this enzyme being able to sialylate surface glycoproteins and consequently masks lactosamine residues by sialylation or by producing sialylated glycan ligands for lectins. Plasma membrane-associated sialyltransferase has also been observed on intestinal cells by immunoelectron microscopy [70]. A surface sialyltransferase and a surface neuraminidase have been implicated in the platelet-platelet interaction that occurs in the first step of hemostasis. Sialyltransferases bind endogenous surface sialic acid acceptors on adjacent cells, thereby forming an adhesive bridge between them. Then, the adhering cells are forced to separate and removal of incorporated sialic acid by neuraminidase activity allows the regeneration of new surface sialic acid acceptors, which enables further aggregation [71].

The presence of sialyltransferases at the cell surface has also been described in parasites, with the trans-sialidase, which exhibits a unique enzymatic activity, having been widely described in trypanosomatids. It is anchored to the cell surface of the *Trypanosoma cruzi* and other parasites via a glycosylphosphatidylinositol linker and promotes transfer of α(2,3)-linked sialic acid from the host's glycoconjugates to acceptor proteins on the parasite's surface or to recipient host molecules. In the absence of suitable acceptors, trans-sialidase also displays neuraminidase activity leading to hydrolysis of α(2,3)-linked sialyl residues [72-74].

Cell Surface Glycolipid Glycosyltransferases

The first evidence regarding glycosylation of glycolipids at the cellular surface came from investigations performed by Hakomori and co-workers [44]. They demonstrated that the GSL linked covalently to artificial supports were glycosylated when NIL and BHK cells contacted this type of complex. The glycosylation occurred under physiological conditions, with or without exogenously added sugar nucleotide, and required no detergents. Experiments performed on mouse 3T3 fibroblasts demonstrated the incorporation of galactose, from extracellular UDP-galactose, into endogenous glycolipid in a cell-contact dependent process [43]. The presence in plasma membranes of a galactosyltransferase active on endogenous lipid-soluble and insoluble membrane-associated acceptors has also been suggested [75]. In addition, the existence of fucosyltransferase activities at the cell surface from both neuron and glial cells was proposed [76]. Ganglioside sialylation has been previously demonstrated in synaptosomal membrane from calf and rat brains and also during the development of neuronal cell cultures [76-78]. Overall, these pioneer experiments suggest the presence of glycolipid glycosyltransferases associated with the outside face of the plasma membrane. However, as there was concern that cross-contamination of biochemically isolated membrane fractions may have occurred, progress in this research area has been limited. More recently, the possible involvement of plasma membrane-associated ST3Gal-V in GM3 synthesis in the thymuses of dexamethasone administered mice has been suggested [79].

Up to this point, the evidence for glycosyltransferases present at the cell surface has been based on indirect experiments in which sugar nucleotides were included in the medium surrounding suspensions or monolayers of presumably intact cells. However, in recent years, the development of new molecular and biochemical approaches has produced a notable growth in information concerning the presence and significance of ganglioside glycosyltransferases and glycohydrolases associated with the outer layer of the plasma membrane. In 2010, our laboratory demonstrated for the first time the expression of ST8Sia-I at the cell surface of Chinese hamster ovary (CHO)-K1 cells (ectopic ST8Sia-I expression) and SK-Mel-28 human melanoma cells (endogenous ST8Sia-I expression) (Figure 3) [28]. This membrane-integrated sialyltransferase was able to sialylate endogenously synthesized GM3 ganglioside as well as exogenously incorporated acceptor substrate. Furthermore, using a specific enzyme-linked immunosorbent assay combined with confocal fluorescence microscopy, it was demonstrated that ecto-ST8Sia-I was able to sialylate hydrophobically or covalently immobilized GM3 onto a solid surface. More interestingly, ecto-ST8Sia-I displayed enzymatic activity on the GM3 substrate belonging to the surface of neighboring cells (trans-catalytic activity) which did not express endogenous ST8Sia-I, with ecto-ST8Sia-I synthesizing GD3 ganglioside at the cell surface by using both the exogenous and endogenous donor substrate (CMP-NeuAc) available at the extracellular milieu (Figure 4). It is worth mentioning that the presence of CMP-NeuAc in human serum has been previously demonstrated [80, 81], and more recently it was shown that thrombin-activated platelets, under physiological conditions, can efficiently and effectively substitute for CMP-NeuAc in extracellular ST6Gal-1 mediated sialylation of target cell surfaces [82]. It has also been demonstrated that trans-catalytic activity of ecto-ST8Sia-I was remarkably reduced when the expression of the substrate was inhibited by using a specific inhibitor of glycolipids biosynthesis, thus revealing the lipidic nature of the acceptor [29]. Extending the previous findings to another cell type, results obtained in Raw264.7 macrophage cell line undoubtedly showed that endogenous ST8Sia-I is present in both the Golgi complex and at the cell surface level (Figure 3, Vilcaes and Daniotti, unpublished results).

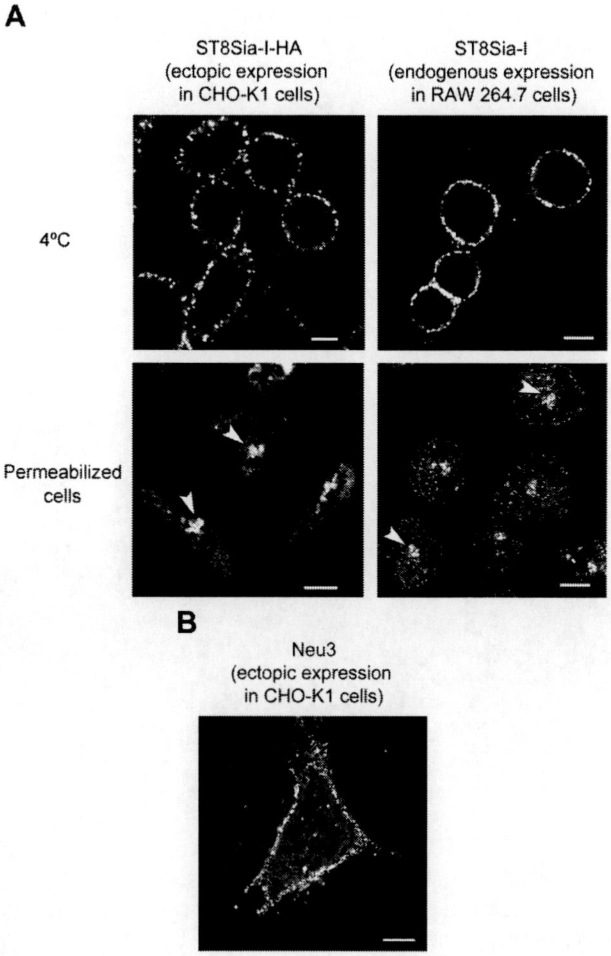

Figure 3. Immunofluorescence detection of ST8Sia-I sialyltransferase and Neu3 sialidase. (A) Left panel, plasma membrane localization of ST8Sia-I-HA detected when CHO-K1 cells stably expressing the enzyme were immunostained with an antibody to HA at 4°C for 45 min, before being fixed and incubated with a secondary antibody conjugated to fluorescent molecule (4°C, upper panel). Alternatively, typical Golgi complex localization detected by immunostaining with antibody to HA epitope previous to fixation and permeabilization (Permeabilized cells, lower panel). Right panel, endogenous detection of ST8Sia-I in Raw264.7 macrophage cell line with an antibody to ST8Sia-I following the same procedure mentioned above. Arrowheads indicate Golgi complex localization of ST8Sia-I. (B) Plasma membrane localization of human Neu3 transiently expressed in CHO-K1 cells detected with antibody to Neu3. In all cases, cells were visualized by confocal microscopy. Single confocal sections were taken every 0.8 μm parallel to the coverslip. Scale bar = 10 μm.

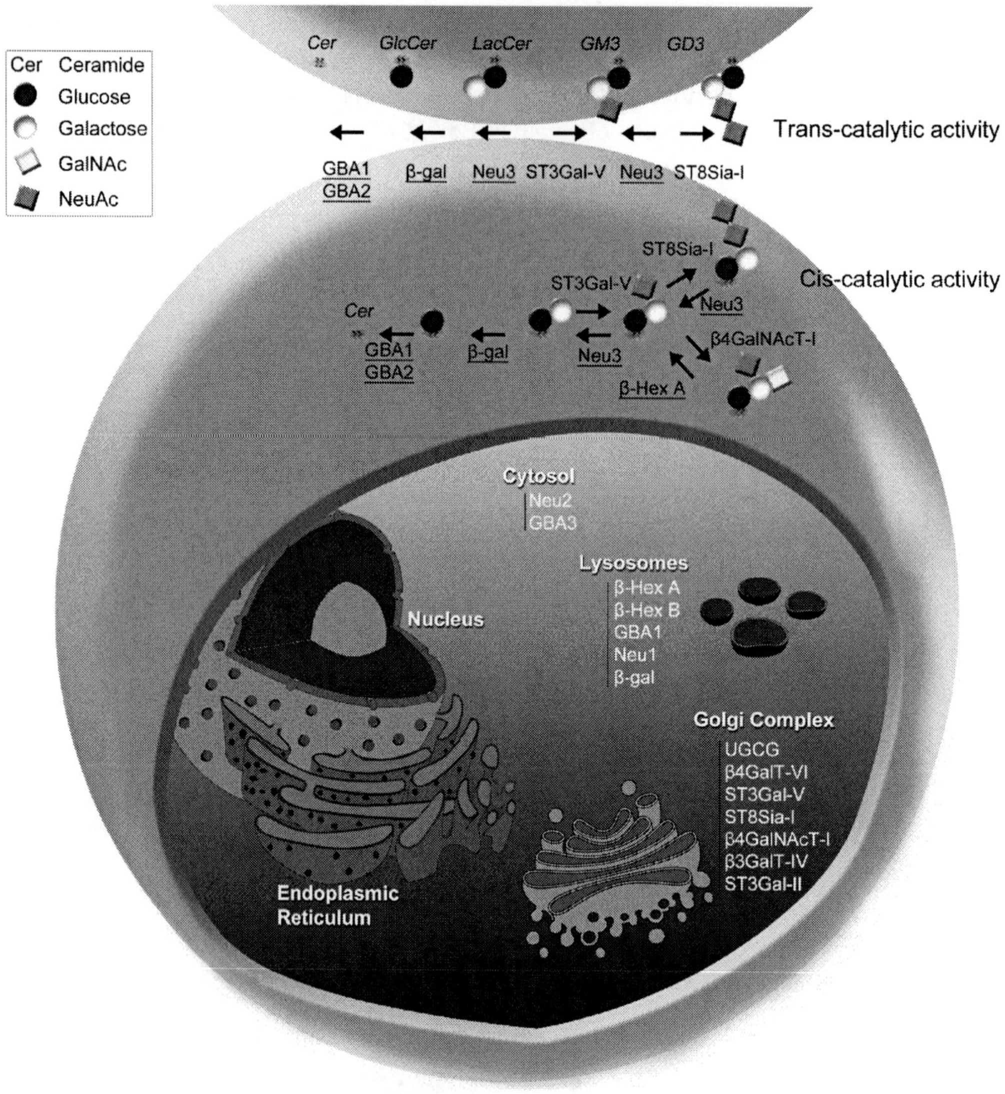

Figure 4. Schematic representation of enzymes and processes associated with GSL metabolism. Intracellular localization of anabolism and catabolism enzymes associated with GSLs (mainly gangliosides), is depicted. At the cell surface, remodeling of GSLs by plasma membrane-associated glycohydrolases (underlined) and glycosyltransferases is represented. As shown in the scheme, GM3 may be catabolized by glycohydrolases, or eventually, be used for GD3 or GM2 biosynthesis by plasma membrane-associated ST8Sia-I and β4GalNAcT-I, respectively (cis-catalytic activity). Eventually, GM3 exposed in membranes from neighboring cells may be modified by glycohydrolases and glycosyltransferases (trans-catalytic activity). β-gal, β-galactosidase.

The expression of another ganglioside glycosyltransferase, β4GalNAcT-I, was also detected at the cell surface of epithelial cells, whose catalytic activity was only observed after feeding the cells with exogenous GM3 substrate [28]. In this context, it is worth mentioning that a considerable fraction of the gangliosides produced by a cell is released to the extracellular milieu, and potentially, can be transferred to the plasma membrane of other cells [83-85]. Although the process of shedding and uptake has been widely documented in both normal and tumor cells, its physiological significance is still unknown. As previously suggested [86], glycolipid composition of the plasma membrane may be remodeled not only by direct uptake from the extracellular environment, but also through later transformation by the plasma membrane-associated enzyme (Figure 3). Although ST3Gal-V, ST3Gal-II and β3GalT-IV enzyme activities could not be detected at the cell surface of epithelial CHO-K1 cells [28], this does not discard their possible presence in other cell types and this still remains to be investigated.

CATABOLISM OF GANGLIOSIDE AT THE CELL SURFACE

The constitutive degradation of gangliosides takes place primarily in lysosomes, as mentioned above. However, increasing evidence suggests that several glycohydrolases are associated with the plasma membrane and that they are involved, together with glycosyltransferases, in plasma membrane GSL remodeling [for reviews, see [87, 88]. To date, the glycohydrolases found to be associated with the cell surface are sialidase Neu3 [89], β-hexosaminidase [90], β-galactosidase and β-glucosidase [91]. In addition, research indicates that the molecular identities of some of the plasma membrane-associated glycohydrolases are different from those found in lysosomes (e.g., plasma membrane bound sialidase Neu3 *vs* lysosomal sialidase Neu2 [89] or plasma membrane-bound β-glucosidase GBA2 *vs* lysosomal β-glucosidase GBA1 [92]. Nevertheless, an immunological and biochemical characterization of the plasma membrane-bound β-hexosaminidase A revealed its lysosomal origin, which suggests that fusion processes between lysosomes and the cell surface may occur [90], but it is still not clear whether plasma membrane-bound and lysosomal β-glucosidase are the same enzyme or not.

β-Glucosidase

β-glucosidases are glycohydrolases that cleave terminal β-linked glucose residues from glycoconjugates, and among the GSLs, β-glucosidases hydrolyze GlcCer to produce ceramide (Figures 2 and 4). Degradation of GlcCer has been mainly studied at the lysosomal level, with GlcCer having been shown to reach the lysosomes through the endosomal pathway and to be degraded in the lysosome lumen by β-glucosidase GBA1. Mutations in the *Gba1* gene in human cause Gaucher disease, a severe lysosomal storage disorder characterized by the accumulation of GlcCer in cells and in certain organs. In addition to the lysosomal GBA1, a distinct β-glucosidase that is also active towards GlcCer was demonstrated in various human tissues and cell types [93], indicating that the enzyme is involved in a non-lysosomal GSL metabolism. In addition, subcellular fractionation analysis has revealed that the glucosidase is not located in lysosomes, but instead in compartments with a considerably lower density. The non-lysosomal glucosidase differs in several respects from its lysosomal counterpart: it is not deficient in Gaucher disease, it differs in its pH optimum and inhibitor sensitivity -GBA1 resulted irreversibly inhibited by conduritol B epoxide (CBE), whereas the non-lysosomal (CBE insensitive β-glucosidase) was not inhibited- and it is not stimulated by the lysosomal activator protein saposin C. When this non-lysosomal β-glucosidase was then cloned, it was identified as GBA2 [92, 94]. Some discrepancy exists with respect to its cellular localization, as it has been reported that GBA2 is present at or close to the cell surface [92] and localizes in the endoplasmic reticulum and Golgi complex [95, 96], with the localization of GBA2 at the Golgi, predominantly in cis compartments, implying that GBA2 activity is in close proximity to the site of GlcCer synthesis. However, the ceramide generated by GBA2 can be rapidly converted into sphingomyelin, suggesting that GBA2 is in relatively close contact with a sphingomyelin synthase [92].

Aureli et al. [91] showed that both the lysosomal GBA1 (or a β-glucosidase isoform with a substrate specificity and sensitivity to inhibitors similar to GBA1) and the non-lysosomal GBA2 are partially associated with the outer layer of the plasma membranes of human fibroblasts, where they can hydrolyze the natural substrate GlcCer and display trans-catalytic activity. Knocking out the non-lysosomal GBA2 in mice results in the accumulation of GlcCer outside the lysosomes in various tissues (testis, brain and liver) [96], with abnormal formation of sperm and decreased fertility. Recently, it has also been demonstrated that the *Gba2* gene is mutated in genetic neurological diseases such as hereditary spastic paraplegia and cerebellar ataxia [97-99]. Nevertheless, the role of GBA2 in neurodegeneration is still not clear.

β-Hexosaminidase

β-Hexosaminidases are glycohydrolases that cleave terminal β-linked *N*-acetyl glucosamine (GlcNAc) or *N*-acetyl galactosamine (GalNAc) residues from glycoconjugates such as oligosaccharides, glycolipids and glycoproteins. These are dimeric enzymes, derived from the assembly of two subunits (α and β). Two major lysosomal isoforms exist in human tissues: β-Hex A (αβ) and β-Hex B (ββ); with the third isoenzyme, β-Hex S (αα), being less abundant [100]. Both the major isoenzymes are able to hydrolyze neutral natural and artificial substrates, but only β-Hex A is able to remove GalNAc from GM2 and GD2 in the presence of GM2 activator protein (Figures 2 and 4). Genetic defects in either of the genes encoding the α and β subunits of β-Hexosaminidases or the GM2 activator protein causes one of the three forms of GM2 gangliosidosis: Tay-Sachs disease, Sandhoff disease and the AB variant of Sandhoff disease, respectively.

The first evidence for the presence of β-hexosaminidase activity associated with the plasma membrane was found in studies of subcellular fractionation of peripheral blood lymphocytes and monocytes from multiple sclerosis patients [101]. Later, this was confirmed by Mencarelli et al. [90], who demonstrated the presence of an active β-Hex A in the plasma membrane from human fibroblasts, which represented 3-5% of total β-hexosaminidase activity. β-Hex A has been found associated to the external leaflet of the plasma membrane and it is also active towards artificial substrates and the natural substrate GM2. The biochemical characterization has indicated that the plasma membrane and lysosomal β-hexosaminidase A are the same enzyme, suggesting that fusion events of lysosomes with the plasma membrane may occur. Discovery of the molecular basis would explain how this enzyme behaves as a soluble protein in lysosomes and membrane associated protein at the cell surface.

Very little is known about the biological meaning of the presence of the enzyme at the cell surface. When relationships between activities of β-glucosidase, β-galactosidase and β-hexosaminidase were investigated in human fibroblast from patients with Gaucher Disease

[102], as expected, the total β-glucosidase activity was much lower in Gaucher disease cell lines than in control fibroblast. However, whereas the total activity of β-galactosidase and β-hexosaminidase was not significantly different in homogenates from Gaucher disease cells compared to control cells, the activities of these enzymes increased at the cell surface. These results demonstrated the existence of a compensatory mechanism among cell surface-associated GSL glycohydrolases, which may serve to maintain membrane properties.

β-Galactosidase

β-galactosidases are hydrolytic enzymes that catalyze the hydrolysis of β-linked galactose residues from glycoconjugates. Substrates of different β-galactosidases include ganglioside GM1, LacCer, lactose and various glycoproteins, with two different β-galactosidases involved in GSL metabolism having been described: one that catalyzes the hydrolysis of galactose from GalCer and LacCer; and one that catalyzes the hydrolysis of the terminal galactose of GM1 (Figures 2 and 4). A deficiency in these β-galactosidases can cause Krabbe disease and GM1 gangliosidosis, respectively.

In addition to the lysosomal β-galactosidase, it has been described that this enzyme is associated with the outer layer of plasma membranes of human fibroblasts and is able to hydrolyze the natural substrate LacCer. Interestingly, it was also found that this hydrolytic enzyme displays a trans-catalytic activity in living cells, by using substrates present in the cell surface of neighboring cells. β-galactosidase activity associated with the cell surface shows a gradual increase during both neuronal cell differentiation and aging, which led to propose that β-galactosidase could be considered a marker for senescence [103, 104]. An important point which is still unclear, however, is whether the enzyme expressed at the cell surface is the same as the one found in lysosomes.

Sialidase

Sialidases hydrolyze the terminal non-reducing sialic acid linkage in glycoproteins, glycolipids, polysaccharides and synthetic molecules (Figure 2). Four types of mammalian sialidases have been identified and characterized to date, designated Neu1, Neu2, Neu3 and Neu4, with each one having different subcellular localization and substrate specificity.

Neu1 is a lysosomal sialidase that hydrolyzes preferentially glycoproteins and oligosaccharides. It is associated with carboxypeptidase protective protein/cathepsin A and β-galactosidase as a complex, with dissociation of the complex leading to sialidase inactivation [105]. Deficiency of Neu1 causes a disease called sialidosis, which is associated with an increased lysosomal exocytosis [38]. Small amounts of Neu1 have been also found in association with the plasma membrane of human erythrocytes in the absence of the protective protein/cathepsin A [106].

Neu2 and Neu4 have broad substrate specificity and can act on glycoproteins and oligosaccharides as well as on glycolipids, although their *in vivo* role is not fully understood. Neu2 is a cytosolic sialidase. Among other important functions, it participates in muscle cell and neuronal differentiation. Neu4 is expressed in two isoforms derived from alternating splicing, and has been found to be present in lysosomes, mitochondria and other subcellular compartments such as the endoplasmic reticulum.

Among human sialidases, the plasma membrane-associated sialidase Neu3 is by far the most studied. It was first cloned from a bovine brain library based on peptide sequence information obtained with the purified enzyme protein [89], and later from the human genome [107, 108]. Neu3 is a key glycosidase for ganglioside degradation that preferentially hydrolyzes the α2-3 and α2-8 external sialyl linkages, but is ineffective on the inner sialic acid residues of GM1 and GM2. Consequently, the gangliosides GM3, GD3, GD1a, GD1b and GT1b, but not GM1 and GM2, are good substrates for the enzyme (see pathway of ganglioside biosynthesis in Figure 1). Neu3 modifies the cell surface ganglioside composition by shifting polysialylated species to the monosialo derivatives and GM3 to LacCer (Figure 4). Besides, sialidase Neu3 is able to hydrolyze the ganglioside substrates present in both the plasma membrane from its own cell or when exposed at the cell surface of neighboring cells, thus supporting its trans-catalytic activity [109].

The major subcellular localization of Neu3 sialidase is at the plasma membrane, but it also occurs in other cellular membrane components such as the endosomal compartments [110]. Because of its association with the plasma membrane, Neu3 is thought to be an important molecule for various cell surface events. Indeed, recent findings have provided evidence that Neu3 is involved not only in the regulation of ganglioside expression, but also in the regulation of transmembrane signaling by interacting with signaling molecules including caveolin-1 [111], β4 integrin [112, 113], epidermal growth factor receptor [114], insulin receptor [115] and β1 integrin [116].

Up-regulation of Neu3 is observed in various types of cancer, including colon, renal, breast, ovarian, melanoma and prostate cancers. Neu3 suppresses apoptosis, promotes cell adhesion and causes an accelerated development of malignant phenotypes. In addition, it has been reported that Neu3 participates in different cellular processes such as differentiation, proliferation and endocytosis [117-120]. We have recently demonstrated that Neu3 is also able to modulate clathrin-mediated endocytosis (Rodriguez-Walker, Vilcaes and Daniotti, unpublished results), delineating a previously unidentified cellular function of Neu3.

The Neu3 knock-out mice did not reveal any accumulation of gangliosides in the brain or colon mucosa, or any obvious abnormalities in their growth, development, behavior, or fertility. However, mice were less susceptible than wild-type mice to the colitis-associated colon carcinogenesis induced by azoxymethane and dextran sodium sulfate, suggesting an important role of Neu3 in inflammation-dependent tumor development [121].

Overall, the results suggest that Neu3 might play a crucial role in regulation of cell surface functions besides ganglioside catabolism.

Conclusion

Current evidence indicates that both ganglioside anabolic and catabolic enzymes are associated with the plasma membrane (Figure 4). This modifies the original and highly compartmentalized concept in which glycohydrolases were thought to be only in lysosomes and glycosyltransferases only at the Golgi complex, involving extra levels of complexity in the overall regulation of GSL expression.

Numerous studies have illustrated how GSLs can interact with several membrane proteins and thereby modulate a number of biological functions. Related to this, it is currently known that gangliosides in particular have a role in cell proliferation and differentiation through the modulation of tyrosine kinase growth factor receptors such as epidermal growth factor receptor [122], insulin receptor [123, 124], TrkA receptor [125, 126] and the platelet-derived growth factor receptor [127, 128]. The coexistence of glycosyltransferases and glycohydrolases (e.g., sialyltransferase and sialidase; β-hexosaminyltransferase and β-hexosaminidase) on the cell surface, support the hypothesis that local changes in GSLs at the plasma membrane could be a means of rapidly modifying the membrane composition and organization, and eventually, modulating cell signaling in response to different stimuli. Otherwise, it would require much more time to neobiosynthesize the GSLs at the Golgi complex and their subsequent intracellular transport to the plasma membrane (Figure 4).

ACKNOWLEDGMENTS

This work was supported in part by Grants from Secretaría de Ciencia y Tecnología (SECyT), Universidad Nacional de Córdoba (UNC), Consejo Nacional de Investigaciones Científicas y Técnicas (CONICET) and Agencia Nacional de Promoción Científica y Tecnológica (ANPCyT).

J. L. D. and A. A. V. are career investigators of CONICET (Argentina). M. R.-W. and F. M. R. are recipients of CONICET fellowships. We apologize to all colleagues whose relevant work could unfortunately not be cited here because of space limitations.

REFERENCES

[1] Maccioni, H. J. (2007) Glycosylation of glycolipids in the Golgi complex. *J. Neurochem.* 103, 81-90.

[2] Daniotti, J. L., and Iglesias-Bartolome, R. (2011) Metabolic pathways and intracellular trafficking of gangliosides. *IUBMB life* 63, 513-520.

[3] Daniotti, J. L., Crespo, P. M., and Yamashita, T. (2006) *In vivo* modulation of epidermal growth factor receptor phosphorylation in mice expressing different gangliosides. *J. Cell Biochem.* 99, 1442-1451.

[4] Hakomori, S., Handa, K., Iwabuchi, K., Yamamura, S., and Prinetti, A. (1998) New insights in glycosphingolipid function: 'glycosignaling domain,' a cell surface assembly of glycosphingolipids with signal transducer molecules, involved in cell adhesion coupled with signaling. *Glycobiology* 8, xi-xix.

[5] Proia, R. L. (2003) Glycosphingolipid functions: insights from engineered mouse models. *Philos. Trans. R. Soc. Lond. B Biol. Sci.* 358, 879-883.

[6] Zurita, A. R., Maccioni, H. J., and Daniotti, J. L. (2001) Modulation of epidermal growth factor receptor phosphorylation by endogenously expressed gangliosides. *Biochem. J.* 355, 465-472.

[7] Cuatrecasas, P. (1973) Gangliosides and membrane receptors for cholera toxin. *Biochemistry* 12, 3558-3566.

[8] Iglesias-Bartolome, R., Trenchi, A., Comin, R., Moyano, A. L., Nores, G. A., and Daniotti, J. L. (2009) Differential endocytic trafficking of neuropathy-associated antibodies to GM1 ganglioside and cholera toxin in epithelial and neural cells. *Biochim. Biophys. Acta* 1788, 2526-2540.

[9] Spooner, R. A., Smith, D. C., Easton, A. J., Roberts, L. M., and Lord, J. M. (2006) Retrograde transport pathways utilised by viruses and protein toxins. *Virol. J.* 3, 26.

[10] Surolia, A., Bachhawat, B. K., and Podder, S. K. (1975) Interaction between lectin from Ricinus communis and liposomes containing gangliosides. *Nature* 257, 802-804.

[11] Nores, G. A., Lardone, R. D., Comin, R., Alaniz, M. E., Moyano, A. L., and Irazoqui, F. J. (2008) Anti-GM1 antibodies as a model of the immune response to self-glycans. *Biochim. Biophys. Acta* 1780, 538-545.

[12] Willison, H. J., and Yuki, N. (2002) Peripheral neuropathies and anti-glycolipid antibodies. *Brain* 125, 2591-2625.

[13] Zhang, X., and Kiechle, F. L. (2004) Review: Glycosphingolipids in health and disease. *Ann. Clin. Lab. Sci.* 34, 3-13.

[14] Daniotti, J. L., Vilcaes, A. A., Torres Demichelis, V., Ruggiero, F. M., and Rodriguez-Walker, M. (2013) Glycosylation of Glycolipids in Cancer: Basis for Development of Novel Therapeutic Approaches. *Front. Oncol.* 3, 306.

[15] Kolter, T. (2012) Ganglioside Biochemistry. *ISRN Biochemistry* 2012, 36.

[16] Kolter, T., and Sandhoff, K. (2010) Lysosomal degradation of membrane lipids. *FEBS Lett.* 584, 1700-1712.

[17] Sandhoff, K., and Harzer, K. (2013) Gangliosides and gangliosidoses: principles of molecular and metabolic pathogenesis. *J. Neurosci.* 33, 10195-10208.

[18] Sandhoff, K., and Kolter, T. (2003) Biosynthesis and degradation of mammalian glycosphingolipids. *Philos. Trans. R. Soc. Lond. B Biol. Sci.* 358, 847-861.

[19] Merrill, A. H., Jr. (2011) Sphingolipid and glycosphingolipid metabolic pathways in the era of sphingolipidomics. *Chemical reviews* 111, 6387-6422.

[20] Tettamanti, G. (2004) Ganglioside/glycosphingolipid turnover: new concepts. *Glycoconj. J.* 20, 301-317.

[21] Hanada, K., Kumagai, K., Yasuda, S., Miura, Y., Kawano, M., Fukasawa, M., and Nishijima, M. (2003) Molecular machinery for non-vesicular trafficking of ceramide. *Nature* 426, 803-809.

[22] Halter, D., Neumann, S., van Dijk, S. M., Wolthoorn, J., de Maziere, A. M., Vieira, O. V., Mattjus, P., Klumperman, J., van Meer, G., and Sprong, H. (2007) Pre- and post-Golgi translocation of glucosylceramide in glycosphingolipid synthesis. *J. Cell Biol.* 179, 101-115.

[23] D'Angelo, G., Vicinanza, M., and De Matteis, M. A. (2008) Lipid-transfer proteins in biosynthetic pathways. *Curr. Opin. Cell Biol.* 20, 360-370.

[24] Lannert, H., Bunning, C., Jeckel, D., and Wieland, F. T. (1994) Lactosylceramide is synthesized in the lumen of the Golgi apparatus. *FEBS Lett.* 342, 91-96.

[25] Crespo, P. M., Iglesias-Bartolome, R., and Daniotti, J. L. (2004) Ganglioside GD3 traffics from the trans-Golgi network to plasma membrane by a Rab11-independent and brefeldin A-insensitive exocytic pathway. *J. Biol. Chem.* 279, 47610-47618.

[26] Iglesias-Bartolome, R., Crespo, P. M., Gomez, G. A., and Daniotti, J. L. (2006) The antibody to GD3 ganglioside, R24, is rapidly endocytosed and recycled to the plasma

membrane via the endocytic recycling compartment. Inhibitory effect of brefeldin A and monensin. *FEBS J.* 273, 1744-1758.

[27] Crespo, P. M., von Muhlinen, N., Iglesias-Bartolome, R., and Daniotti, J. L. (2008) Complex gangliosides are apically sorted in polarized MDCK cells and internalized by clathrin-independent endocytosis. *FEBS J.* 275, 6043-6056.

[28] Crespo, P. M., Demichelis, V. T., and Daniotti, J. L. (2010) Neobiosynthesis of glycosphingolipids by plasma membrane-associated glycosyltransferases. *J. Biol. Chem.* 285, 29179-29190.

[29] Vilcaes, A. A., Demichelis, V. T., and Daniotti, J. L. (2011) Trans-activity of plasma membrane-associated ganglioside sialyltransferase in mammalian cells. *J. Biol. Chem.* 286, 31437-31446.

[30] Boccuto, L., Aoki, K., Flanagan-Steet, H., Chen, C. F., Fan, X., Bartel, F., Petukh, M., Pittman, A., Saul, R., Chaubey, A., Alexov, E., Tiemeyer, M., Steet, R., and Schwartz, C. E. (2014) A mutation in a ganglioside biosynthetic enzyme, ST3GAL5, results in salt & pepper syndrome, a neurocutaneous disorder with altered glycolipid and glycoprotein glycosylation. *Hum. Mol. Genet.* 23, 418-433.

[31] Fragaki, K., Ait-El-Mkadem, S., Chaussenot, A., Gire, C., Mengual, R., Bonesso, L., Beneteau, M., Ricci, J. E., Desquiret-Dumas, V., Procaccio, V., Rotig, A., and Paquis-Flucklinger, V. (2013) Refractory epilepsy and mitochondrial dysfunction due to GM3 synthase deficiency. *Eur. J. Hum. Genet.* 21, 528-534.

[32] Simpson, M. A., Cross, H., Proukakis, C., Priestman, D. A., Neville, D. C., Reinkensmeier, G., Wang, H., Wiznitzer, M., Gurtz, K., Verganelaki, A., Pryde, A., Patton, M. A., Dwek, R. A., Butters, T. D., Platt, F. M., and Crosby, A. H. (2004) Infantile-onset symptomatic epilepsy syndrome caused by a homozygous loss-of-function mutation of GM3 synthase. *Nat. Genet.* 36, 1225-1229.

[33] Boukhris, A., Schule, R., Loureiro, J. L., Lourenco, C. M., Mundwiller, E., Gonzalez, M. A., Charles, P., Gauthier, J., Rekik, I., Acosta Lebrigio, R. F., Gaussen, M., Speziani, F., Ferbert, A., Feki, I., Caballero-Oteyza, A., Dionne-Laporte, A., Amri, M., Noreau, A., Forlani, S., Cruz, V. T., Mochel, F., Coutinho, P., Dion, P., Mhiri, C., Schols, L., Pouget, J., Darios, F., Rouleau, G. A., Marques, W., Jr., Brice, A., Durr, A., Zuchner, S., and Stevanin, G. (2013) Alteration of ganglioside biosynthesis responsible for complex hereditary spastic paraplegia. *Am. J. Hum. Genet.* 93, 118-123.

[34] Harlalka, G. V., Lehman, A., Chioza, B., Baple, E. L., Maroofian, R., Cross, H., Sreekantan-Nair, A., Priestman, D. A., Al-Turki, S., McEntagart, M. E., Proukakis, C., Royle, L., Kozak, R. P., Bastaki, L., Patton, M., Wagner, K., Coblentz, R., Price, J., Mezei, M., Schlade-Bartusiak, K., Platt, F. M., Hurles, M. E., and Crosby, A. H. (2013) Mutations in B4GALNT1 (GM2 synthase) underlie a new disorder of ganglioside biosynthesis. *Brain* 136, 3618-3624.

[35] Hu, H., Eggers, K., Chen, W., Garshasbi, M., Motazacker, M. M., Wrogemann, K., Kahrizi, K., Tzschach, A., Hosseini, M., Bahman, I., Hucho, T., Muhlenhoff, M., Gerardy-Schahn, R., Najmabadi, H., Ropers, H. H., and Kuss, A. W. (2011) ST3GAL3 mutations impair the development of higher cognitive functions. *Am. J. Hum. Genet.* 89, 407-414.

[36] Edvardson, S., Baumann, A. M., Muhlenhoff, M., Stephan, O., Kuss, A. W., Shaag, A., He, L., Zenvirt, S., Tanzi, R., Gerardy-Schahn, R., and Elpeleg, O. (2013) West syndrome caused by ST3Gal-III deficiency. *Epilepsia* 54, e24-27.

[37] Dawkins, J. L., Hulme, D. J., Brahmbhatt, S. B., Auer-Grumbach, M., and Nicholson, G. A. (2001) Mutations in SPTLC1, encoding serine palmitoyltransferase, long chain base subunit-1, cause hereditary sensory neuropathy type I. *Nat. Genet.* 27, 309-312.
[38] d'Azzo, A., and Bonten, E. (2010) Molecular mechanisms of pathogenesis in a glycosphingolipid and a glycoprotein storage disease. *Biochem. Soc. Trans.* 38, 1453-1457.
[39] Kolter, T., and Sandhoff, K. (2006) Sphingolipid metabolism diseases. *Biochim. Biophys. Acta* 1758, 2057-2079.
[40] Platt, F. M. (2014) Sphingolipid lysosomal storage disorders. *Nature* 510, 68-75.
[41] Patt, L. M., and Grimes, W. J. (1974) Cell surface glycolipid and glycoprotein glycosyltransferases of normal and transformed cells. *J. Biol. Chem.* 249, 4157-4165.
[42] Roth, S., McGuire, E. J., and Roseman, S. (1971) Evidence for cell-surface glycosyltransferases. Their potential role in cellular recognition. *J. Cell Biol.* 51, 536-547.
[43] Roth, S., and White, D. (1972) Intercellular contact and cell-surface galactosyl transferase activity (cell culture-mouse-radioautography-contact inhibition-cis-and trans-galactosylation). *Proc. Natl. Acad. Sci. U.S.A.* 69, 485-489.
[44] Yogeeswaran, G., Laine, R. A., and Hakomori, S. (1974) Mechanism of cell contact-dependent glycolipid synthesis: further studies with glycolipid-glass complex. *Biochem. Biophys. Res. Commun.* 59, 591-599.
[45] Roth, J., and Berger, E. G. (1982) Immunocytochemical localization of galactosyltransferase in HeLa cells: codistribution with thiamine pyrophosphatase in trans-Golgi cisternae. *J. Cell Biol.* 93, 223-229.
[46] Slot, J. W., and Geuze, H. J. (1983) Immunoelectron microscopic exploration of the Golgi complex. *J. Histochem. Cytochem.* 31, 1049-1056.
[47] Mrkoci Felner, K., Niederer, E., and Berger, E. G. (1997) Flow cytometric detection of the Golgi apparatus using antibodies to glycosyltransferases. *Cytometry* 28, 50-57.
[48] Roseman, S. (1970) The synthesis of complex carbohydrates by multiglycosyl-transferase systems and their potential function in intercellular adhesion. *Chem. Phys. Lipids* 5, 270-297.
[49] Shur, B. D., and Roth, S. (1975) Cell surface glycosyltransferases. *Biochim. Biophys. Acta* 415, 473-512.
[50] Suganuma, T., Muramatsu, H., Muramatsu, T., Ihida, K., Kawano, J., and Murata, F. (1991) Subcellular localization of N-acetylglucosaminide beta 1,4-galactosyltransferase revealed by immunoelectron microscopy. *J. Histochem. Cytochem.* 39, 299-309.
[51] Shaper, J. H., Hollis, G. F., and Shaper, N. L. (1988) Evidence for two forms of murine beta-1,4-galactosyltransferase based on cloning studies. *Biochimie* 70, 1683-1688.
[52] Evans, S. C., Lopez, L. C., and Shur, B. D. (1993) Dominant negative mutation in cell surface beta 1,4-galactosyltransferase inhibits cell-cell and cell-matrix interactions. *J. Cell Biol.* 120, 1045-1057.
[53] Youakim, A., Hathaway, H. J., Miller, D. J., Gong, X., and Shur, B. D. (1994) Overexpressing sperm surface beta 1,4-galactosyltransferase in transgenic mice affects multiple aspects of sperm-egg interactions. *J. Cell Biol.* 126, 1573-1583.
[54] Yamamoto, K., Kawakami, H., Sugiyama, M., Atomi, Y., and Hirano, H. (1999) Immunohistochemical localization of beta-1,4-galactosyltransferase in human pancreatic tissues. *Pancreas* 19, 137-142.

[55] Nguyen, T. T., Hinton, D. A., and Shur, B. D. (1994) Expressing murine beta 1,4-galactosyltransferase in HeLa cells produces a cell surface galactosyltransferase-dependent phenotype. *J. Biol. Chem.* 269, 28000-28009.

[56] Runyan, R. B., Versalovic, J., and Shur, B. D. (1988) Functionally distinct laminin receptors mediate cell adhesion and spreading: the requirement for surface galactosyltransferase in cell spreading. *J. Cell Biol.* 107, 1863-1871.

[57] Gong, X., Dubois, D. H., Miller, D. J., and Shur, B. D. (1995) Activation of a G protein complex by aggregation of beta-1,4-galactosyltransferase on the surface of sperm. *Science* 269, 1718-1721.

[58] Macek, M. B., Lopez, L. C., and Shur, B. D. (1991) Aggregation of beta-1,4-galactosyltransferase on mouse sperm induces the acrosome reaction. *Dev. Biol.* 147, 440-444.

[59] Shi, X., Amindari, S., Paruchuru, K., Skalla, D., Burkin, H., Shur, B. D., and Miller, D. J. (2001) Cell surface beta-1,4-galactosyltransferase-I activates G protein-dependent exocytotic signaling. *Development* 128, 645-654.

[60] Lopez, L. C., Bayna, E. M., Litoff, D., Shaper, N. L., Shaper, J. H., and Shur, B. D. (1985) Receptor function of mouse sperm surface galactosyltransferase during fertilization. *J. Cell Biol.* 101, 1501-1510.

[61] Nixon, B., Lu, Q., Wassler, M. J., Foote, C. I., Ensslin, M. A., and Shur, B. D. (2001) Galactosyltransferase function during mammalian fertilization. *Cells Tissues Organs* 168, 46-57.

[62] Shur, B. D., and Neely, C. A. (1988) Plasma membrane association, purification, and partial characterization of mouse sperm beta 1,4-galactosyltransferase. *J. Biol. Chem.* 263, 17706-17714.

[63] Tengowski, M. W., Wassler, M. J., Shur, B. D., and Schatten, G. (2001) Subcellular localization of beta 1,4-galactosyltransferase on bull sperm and its function during sperm-egg interactions. *Mol. Reprod. Dev.* 58, 236-244.

[64] Barber, A. J., and Jamieson, G. A. (1971) Platelet collagen adhesion characterization of collagen glucosyltransferase of plasma membranes of human blood platelets. *Biochim. Biophys. Acta* 252, 533-545.

[65] Bosmann, H. B. (1971) Platelet adhesiveness and aggregation: the collagen:glycosyl, polypeptide:N-acetylgalactosaminyl and glycoprote- in:galactosyl transferases of human platelets. *Biochem. Biophys. Res. Commun.* 43, 1118-1124.

[66] Jamieson, G. A., Fuller, N. A., Barber, A. J., and Lombart, C. (1971) Membrane glycoproteins of human blood platelets. *Ser Haematol.* 4, 125-134.

[67] Lombart, C., Okumura, T., and Jamieson, G. A. (1974) Isolation of a surface glycoprotein of human platelets. *FEBS Lett.* 41, 30-34.

[68] Gross, H. J., Merling, A., Moldenhauer, G., and Schwartz-Albiez, R. (1996) Ecto-sialyltransferase of human B lymphocytes reconstitutes differentiation markers in the presence of exogenous CMP-N-acetyl neuraminic acid. *Blood* 87, 5113-5126.

[69] Schwartz-Albiez, R., Merling, A., Martin, S., Haas, R., and Gross, H. J. (2004) Cell surface sialylation and ecto-sialyltransferase activity of human CD34 progenitors from peripheral blood and bone marrow. *Glycoconj. J.* 21, 451-459.

[70] Taatjes, D. J., Roth, J., Weinstein, J., and Paulson, J. C. (1988) Post-Golgi apparatus localization and regional expression of rat intestinal sialyltransferase detected by

immunoelectron microscopy with polypeptide epitope-purified antibody. *J. Biol. Chem.* 263, 6302-6309.

[71] Bosmann, H. B. (1972) Platelet adhesiveness and aggregation. II. Surface sialic acid, glycoprotein: N-acetylneuraminic acid transferase, and neuraminidase of human blood platelets. *Biochim. Biophys. Acta* 279, 456-474.

[72] Frasch, A. C. (2000) Functional diversity in the trans-sialidase and mucin families in Trypanosoma cruzi. *Parasitol. Today* 16, 282-286.

[73] Mucci, J., Risso, M. G., Leguizamon, M. S., Frasch, A. C., and Campetella, O. (2006) The trans-sialidase from Trypanosoma cruzi triggers apoptosis by target cell sialylation. *Cell Microbiol.* 8, 1086-1095.

[74] Muia, R. P., Yu, H., Prescher, J. A., Hellman, U., Chen, X., Bertozzi, C. R., and Campetella, O. (2010) Identification of glycoproteins targeted by Trypanosoma cruzi trans-sialidase, a virulence factor that disturbs lymphocyte glycosylation. *Glycobiology* 20, 833-842.

[75] LaMont, J. T., Gammon, M. T., and Isselbacher, K. J. (1977) Cell-surface glycosyltransferases in cultured fibroblasts: increased activity and release during serum stimulation of growth. *Proc. Natl. Acad. Sci. U.S.A.* 74, 1086-1090.

[76] Matsui, Y., Lombard, D., Massarelli, R., Mandel, P., and Dreyfus, H. (1986) Surface glycosyltransferase activities during development of neuronal cell cultures. *J. Neurochem.* 46, 144-150.

[77] Durrie, R., Saito, M., and Rosenberg, A. (1988) Endogenous glycosphingolipid acceptor specificity of sialosyltransferase systems in intact Golgi membranes, synaptosomes, and synaptic plasma membranes from rat brain. *Biochemistry* 27, 3759-3764.

[78] Preti, A., Fiorilli, A., Lombardo, A., Caimi, L., and Tettamanti, G. (1980) Occurrence of sialyltransferase activity in the synaptosomal membranes prepared from calf brain cortex. *J. Neurochem.* 35, 281-296.

[79] Iwamori, M., and Iwamori, Y. (2005) Changes in the glycolipid composition and characteristic activation of GM3 synthase in the thymus of mouse after administration of dexamethasone. *Glycoconj. J.* 22, 119-126.

[80] Parsons, N. J., Ashton, P. R., Constantinidou, C., Cole, J. A., and Smith, H. (1993) Identification by mass spectrometry of CMP-NANA in diffusible material released from high M(r) blood cell fractions that confers serum resistance on gonococci. *Microb. Pathog.* 14, 329-335.

[81] Smith, H., Parsons, N. J., and Cole, J. A. (1995) Sialylation of neisserial lipopolysaccharide: a major influence on pathogenicity. *Microb. Pathog.* 19, 365-377.

[82] Lee, M. M., Nasirikenari, M., Manhardt, C. T., Ashline, D. J., Hanneman, A. J., Reinhold, V. N., and Lau, J. T. (2014) Platelets support extracellular sialylation by supplying the sugar donor substrate. *J. Biol. Chem.* 289, 8742-8748.

[83] Chang, F., Li, R., and Ladisch, S. (1997) Shedding of gangliosides by human medulloblastoma cells. *Exp. Cell Res.* 234, 341-346.

[84] Kong, Y., Li, R., and Ladisch, S. (1998) Natural forms of shed tumor gangliosides. *Biochim. Biophys. Acta* 1394, 43-56.

[85] Ziulkoski, A. L., Andrade, C. M., Crespo, P. M., Sisti, E., Trindade, V. M., Daniotti, J. L., Guma, F. C., and Borojevic, R. (2006) Gangliosides of myelosupportive stroma

cells are transferred to myeloid progenitors and are required for their survival and proliferation. *Biochem. J.* 394, 1-9.

[86] Aureli, M., Loberto, N., Lanteri, P., Chigorno, V., Prinetti, A., and Sonnino, S. (2011) Cell surface sphingolipid glycohydrolases in neuronal differentiation and aging in culture. *J. Neurochem.* 116, 891-899.

[87] Prinetti, A., Chigorno, V., Mauri, L., Loberto, N., and Sonnino, S. (2007) Modulation of cell functions by glycosphingolipid metabolic remodeling in the plasma membrane. *J. Neurochem.* 103, 113-125.

[88] Sonnino, S., Aureli, M., Loberto, N., Chigorno, V., and Prinetti, A. (2010) Fine tuning of cell functions through remodeling of glycosphingolipids by plasma membrane-associated glycohydrolases. *FEBS Lett.* 584, 1914-1922.

[89] Miyagi, T., Wada, T., Iwamatsu, A., Hata, K., Yoshikawa, Y., Tokuyama, S., and Sawada, M. (1999) Molecular cloning and characterization of a plasma membrane-associated sialidase specific for gangliosides. *J. Biol. Chem.* 274, 5004-5011.

[90] Mencarelli, S., Cavalieri, C., Magini, A., Tancini, B., Basso, L., Lemansky, P., Hasilik, A., Li, Y. T., Chigorno, V., Orlacchio, A., Emiliani, C., and Sonnino, S. (2005) Identification of plasma membrane associated mature beta-hexosaminidase A, active towards GM2 ganglioside, in human fibroblasts. *FEBS Lett.* 579, 5501-5506.

[91] Aureli, M., Masilamani, A. P., Illuzzi, G., Loberto, N., Scandroglio, F., Prinetti, A., Chigorno, V., and Sonnino, S. (2009) Activity of plasma membrane beta-galactosidase and beta-glucosidase. *FEBS Lett.* 583, 2469-2473.

[92] Boot, R. G., Verhoek, M., Donker-Koopman, W., Strijland, A., van Marle, J., Overkleeft, H. S., Wennekes, T., and Aerts, J. M. (2007) Identification of the non-lysosomal glucosylceramidase as beta-glucosidase 2. *J. Biol. Chem.* 282, 1305-1312.

[93] van Weely, S., Brandsma, M., Strijland, A., Tager, J. M., and Aerts, J. M. (1993) Demonstration of the existence of a second, non-lysosomal glucocerebrosidase that is not deficient in Gaucher disease. *Biochim. Biophys. Acta* 1181, 55-62.

[94] Matern, H., Boermans, H., Lottspeich, F., and Matern, S. (2001) Molecular cloning and expression of human bile acid beta-glucosidase. *J. Biol. Chem.* 276, 37929-37933.

[95] Körschen, H. G., Yildiz, Y., Raju, D. N., Schonauer, S., Bönigk, W., Jansen, V., Kremmer, E., Kaupp, U. B., and Wachten, D. (2013) The non-lysosomal β-glucosidase GBA2 is a non-integral membrane-associated protein at the endoplasmic reticulum (ER) and Golgi. *J. Biol. Chem.* 288, 3381-3393.

[96] Yildiz, Y., Matern, H., Thompson, B., Allegood, J. C., Warren, R. L., Ramirez, D. M., Hammer, R. E., Hamra, F. K., Matern, S., and Russell, D. W. (2006) Mutation of beta-glucosidase 2 causes glycolipid storage disease and impaired male fertility. *J. Clin. Invest.* 116, 2985-2994.

[97] Citterio, A., Arnoldi, A., Panzeri, E., D'Angelo, M. G., Filosto, M., Dilena, R., Arrigoni, F., Castelli, M., Maghini, C., Germiniasi, C., Menni, F., Martinuzzi, A., Bresolin, N., and Bassi, M. T. (2014) Mutations in CYP2U1, DDHD2 and GBA2 genes are rare causes of complicated forms of hereditary spastic paraparesis. *J. Neurol.* 261, 373-381.

[98] Martin, E., Schüle, R., Smets, K., Rastetter, A., Boukhris, A., Loureiro, J. L., Gonzalez, M. A., Mundwiller, E., Deconinck, T., Wessner, M., Jornea, L., Oteyza, A. C., Durr, A., Martin, J. J., Schöls, L., Mhiri, C., Lamari, F., Züchner, S., De Jonghe, P., Kabashi, E., Brice, A., and Stevanin, G. (2013) Loss of function of glucocerebrosidase GBA2 is

responsible for motor neuron defects in hereditary spastic paraplegia. *Am. J. Hum. Genet.* 92, 238-244.

[99] Votsi, C., Zamba-Papanicolaou, E., Middleton, L. T., Pantzaris, M., and Christodoulou, K. (2014) A novel GBA2 gene missense mutation in spastic ataxia. *Ann. Hum. Genet.* 78, 13-22.

[100] Sandhoff, K. (2012) My journey into the world of sphingolipids and sphingolipidoses. *Proc. Jpn. Acad. Ser. B Phys. Biol. Sci.* 88, 554-582.

[101] Orlacchio, A., Martino, S., Sarchielli, P., Gallai, V., and Emiliani, C. (1998) Beta-N-acetylhexosaminidase in peripheral blood lymphocytes and monocytes in the different forms and stages of multiple sclerosis. *J. Neurochem.* 71, 1168-1176.

[102] Aureli, M., Bassi, R., Loberto, N., Regis, S., Prinetti, A., Chigorno, V., Aerts, J. M., Boot, R. G., Filocamo, M., and Sonnino, S. (2012) Cell surface associated glycohydrolases in normal and Gaucher disease fibroblasts. *J. Inherit. Metab. Dis.* 35, 1081-1091.

[103] Dimri, G. P., Lee, X., Basile, G., Acosta, M., Scott, G., Roskelley, C., Medrano, E. E., Linskens, M., Rubelj, I., Pereira-Smith, O., and et al. (1995) A biomarker that identifies senescent human cells in culture and in aging skin *in vivo*. *Proc. Natl. Acad. Sci. U.S.A.* 92, 9363-9367.

[104] Geng, Y. Q., Guan, J. T., Xu, X. H., and Fu, Y. C. (2010) Senescence-associated beta-galactosidase activity expression in aging hippocampal neurons. *Biochem. Biophys. Res. Commun.* 396, 866-869.

[105] Galjart, N. J., Gillemans, N., Harris, A., van der Horst, G. T., Verheijen, F. W., Galjaard, H., and d'Azzo, A. (1988) Expression of cDNA encoding the human 'protective protein' associated with lysosomal beta-galactosidase and neuraminidase: homology to yeast proteases. *Cell* 54, 755-764.

[106] D'Avila, F., Tringali, C., Papini, N., Anastasia, L., Croci, G., Massaccesi, L., Monti, E., Tettamanti, G., and Venerando, B. (2013) Identification of lysosomal sialidase NEU1 and plasma membrane sialidase NEU3 in human erythrocytes. *J. Cell Biochem.* 114, 204-211.

[107] Monti, E., Bassi, M. T., Papini, N., Riboni, M., Manzoni, M., Venerando, B., Croci, G., Preti, A., Ballabio, A., Tettamanti, G., and Borsani, G. (2000) Identification and expression of NEU3, a novel human sialidase associated to the plasma membrane. *Biochem. J.* 349, 343-351.

[108] Wada, T., Yoshikawa, Y., Tokuyama, S., Kuwabara, M., Akita, H., and Miyagi, T. (1999) Cloning, expression, and chromosomal mapping of a human ganglioside sialidase. *Biochem. Biophys. Res. Commun.* 261, 21-27.

[109] Papini, N., Anastasia, L., Tringali, C., Croci, G., Bresciani, R., Yamaguchi, K., Miyagi, T., Preti, A., Prinetti, A., Prioni, S., Sonnino, S., Tettamanti, G., Venerando, B., and Monti, E. (2004) The plasma membrane-associated sialidase MmNEU3 modifies the ganglioside pattern of adjacent cells supporting its involvement in cell-to-cell interactions. *J. Biol. Chem.* 279, 16989-16995.

[110] Zanchetti, G., Colombi, P., Manzoni, M., Anastasia, L., Caimi, L., Borsani, G., Venerando, B., Tettamanti, G., Preti, A., Monti, E., and Bresciani, R. (2007) Sialidase NEU3 is a peripheral membrane protein localized on the cell surface and in endosomal structures. *Biochem. J.* 408, 211-219.

[111] Wang, Y., Yamaguchi, K., Wada, T., Hata, K., Zhao, X., Fujimoto, T., and Miyagi, T. (2002) A close association of the ganglioside-specific sialidase Neu3 with caveolin in membrane microdomains. *J. Biol. Chem.* 277, 26252-26259.

[112] Bonardi, D., Papini, N., Pasini, M., Dileo, L., Orizio, F., Monti, E., Caimi, L., Venerando, B., and Bresciani, R. (2014) Sialidase NEU3 dynamically associates to different membrane domains specifically modifying their ganglioside pattern and triggering Akt phosphorylation. *PLoS One* 9, e99405.

[113] Kato, K., Shiga, K., Yamaguchi, K., Hata, K., Kobayashi, T., Miyazaki, K., Saijo, S., and Miyagi, T. (2006) Plasma-membrane-associated sialidase (NEU3) differentially regulates integrin-mediated cell proliferation through laminin- and fibronectin-derived signalling. *Biochem. J.* 394, 647-656.

[114] Wada, T., Hata, K., Yamaguchi, K., Shiozaki, K., Koseki, K., Moriya, S., and Miyagi, T. (2007) A crucial role of plasma membrane-associated sialidase in the survival of human cancer cells. *Oncogene* 26, 2483-2490

[115] Sasaki, A., Hata, K., Suzuki, S., Sawada, M., Wada, T., Yamaguchi, K., Obinata, M., Tateno, H., Suzuki, H., and Miyagi, T. (2003) Overexpression of plasma membrane-associated sialidase attenuates insulin signaling in transgenic mice. *J. Biol. Chem.* 278, 27896-27902.

[116] Tringali, C., Lupo, B., Silvestri, I., Papini, N., Anastasia, L., Tettamanti, G., and Venerando, B. (2012) The plasma membrane sialidase NEU3 regulates the malignancy of renal carcinoma cells by controlling β1 integrin internalization and recycling. *J. Biol. Chem.* 287, 42835-42845.

[117] Anastasia, L., Papini, N., Colazzo, F., Palazzolo, G., Tringali, C., Dileo, L., Piccoli, M., Conforti, E., Sitzia, C., Monti, E., Sampaolesi, M., Tettamanti, G., and Venerando, B. (2008) NEU3 sialidase strictly modulates GM3 levels in skeletal myoblasts C2C12 thus favoring their differentiation and protecting them from apoptosis. *J. Biol. Chem.* 283, 36265-36271.

[118] Kakugawa, Y., Wada, T., Yamaguchi, K., Yamanami, H., Ouchi, K., Sato, I., and Miyagi, T. (2002) Up-regulation of plasma membrane-associated ganglioside sialidase (Neu3) in human colon cancer and its involvement in apoptosis suppression. *Proc. Natl. Acad. Sci. U.S.A.* 99, 10718-10723.

[119] Miyagi, T., and Yamaguchi, K. (2012) Mammalian sialidases: physiological and pathological roles in cellular functions. *Glycobiology* 22, 880-896.

[120] Rodriguez, J. A., Piddini, E., Hasegawa, T., Miyagi, T., and Dotti, C. G. (2001) Plasma membrane ganglioside sialidase regulates axonal growth and regeneration in hippocampal neurons in culture. *J. Neurosci.* 21, 8387-8395.

[121] Yamaguchi, K., Shiozaki, K., Moriya, S., Koseki, K., Wada, T., Tateno, H., Sato, I., Asano, M., Iwakura, Y., and Miyagi, T. (2012) Reduced susceptibility to colitis-associated colon carcinogenesis in mice lacking plasma membrane-associated sialidase. *PLoS One* 7, e41132.

[122] Miljan, E. A., Meuillet, E. J., Mania-Farnell, B., George, D., Yamamoto, H., Simon, H. G., and Bremer, E. G. (2002) Interaction of the extracellular domain of the epidermal growth factor receptor with gangliosides. *J. Biol. Chem.* 277, 10108-10113.

[123] Allende, M. L., and Proia, R. L. (2002) Lubricating cell signaling pathways with gangliosides. *Curr. Opin. Struct. Biol.* 12, 587-592.

[124] Yamashita, T., Hashiramoto, A., Haluzik, M., Mizukami, H., Beck, S., Norton, A., Kono, M., Tsuji, S., Daniotti, J. L., Werth, N., Sandhoff, R., Sandhoff, K., and Proia, R. L. (2003) Enhanced insulin sensitivity in mice lacking ganglioside GM3. *Proc. Natl. Acad. Sci. U.S.A.* 100, 3445-3449.

[125] Fukumoto, S., Mutoh, T., Hasegawa, T., Miyazaki, H., Okada, M., Goto, G., Furukawa, K., and Urano, T. (2000) GD3 synthase gene expression in PC12 cells results in the continuous activation of TrkA and ERK1/2 and enhanced proliferation. *J. Biol. Chem.* 275, 5832-5838.

[126] Mutoh, T., Tokuda, A., Miyadai, T., Hamaguchi, M., and Fujiki, N. (1995) Ganglioside GM1 binds to the Trk protein and regulates receptor function. *Proc. Natl. Acad. Sci. U.S.A.* 92, 5087-5091

[127] Mitsuda, T., Furukawa, K., Fukumoto, S., Miyazaki, H., and Urano, T. (2002) Overexpression of ganglioside GM1 results in the dispersion of platelet-derived growth factor receptor from glycolipid-enriched microdomains and in the suppression of cell growth signals. *J. Biol. Chem.* 277, 11239-11246.

[128] Yates, A. J., Saqr, H. E., and Van Brocklyn, J. (1995) Ganglioside modulation of the PDGF receptor. A model for ganglioside functions. *J. Neurooncol.* 24, 65-73.

[129] Svennerholm, L. (1980) Ganglioside designation. *Adv. Exp. Med. Biol.* 125, 11.

Reviewed by:

Dr. Hugo J. F. Maccioni, CIQUIBIC, Facultad de Ciencias Químicas, Universidad Nacional de Córdoba, Córdoba, Argentina.

Dr. Gustavo Nores, CIQUIBIC, Facultad de Ciencias Químicas, Universidad Nacional de Córdoba, Córdoba, Argentina.

Chapter 5

UNRAVELING THE SPECIFIC ROLE OF SPHINGOMYELINS IN CELL PHYSIOLOGY: WHERE ARE WE?

Joan Bestard-Escalas[1], María Dolores Ledesma[2], Makoto Taniguchi[3], Toshiro Okazaki[3,4], María Laura Martin[5] and Gwendolyn Barceló-Coblijn[1,*]

[1]Research Unit, Hospital Universitari Son Espases, Institut d'Investigació Sanitària de Palma (IdISPa), Palma, Balearic Islands, Spain

[2]Department of Neurobiology, Centro Biología Molecular Severo Ochoa, CSIC-UAM, Madrid, Spain

[3]Medical Research Institute, Kanazawa Medical University, Ishikawa, Japan

[4]Department of Hematology/Immunology, Faculty of Medicine, Kanazawa Medical University, Uchinada, Ishikawa, Japan

[5]Laboratory of Signal Transduction, Memorial Sloan-Kettering Cancer Center, New York, NY, US

ABSTRACT

Sphingomyelin (SM) is one of the most abundant components of cellular membranes in organisms ranging from mammals to protozoa accounting for 10 to 20% of total phospholipid content. It is preferentially concentrated at the plasma membrane outer leaflet and its high packing density and affinity for sterols provide a rigid barrier to the extracellular environment and play a role in the formation of lipid rafts. SM consists of an sphingoid backbone linked to a fatty acid and to phosphocholine moiety. When analyzing SM molecular composition up to four different sphingoid bases and more than 15 different fatty acids may be found, the combination of which account for more than hundred different SM molecular species.

* Corresponding author: Gwendolyn Barceló-Coblijn. Research Unit, Hospital Universitari Son Espases, Institut d'Investigació Sanitària de Palma (IdISPa), Ctra. Valldemossa 79, E-07120 Palma, Balearic Islands, Spain.

A recent study showing the direct interaction between one specific SM molecular species (SMd18:1-18:0) and a protein involved in COPI-transport (p24) indicates that molecular species of a particular lipid class may be as important as the different protein isoforms. Therefore, it is critical to accurately determine the lipid composition when the function of a lipid class is studied. The rapid advances in mass spectrometry applied to lipidomics will definitively help to tailor the specific role that each of the different SM may have in cell physiology.

SM has been traditionally viewed as a simple structural building block of the plasma membrane. Currently, it is accepted that in addition to its structural functions, SM serves as source of important signaling molecules. However, the possibility that SM by itself could be important in regulating crucial cell events has been clearly underestimated despite the existing evidence. Thus, phorbol ester-induced differentiation of HL-60 human promyelocytic leukemia cells into macrophages is coupled to an increase in SM synthesis via Sphingomyelin Synthase (SMS). Conversely, SM hydrolysis induces proliferation of quiescent 3T3 fibroblasts. Similarly, quiescent satellite cells (skeletal muscle stem cells) contain higher SM levels than proliferating satellite cells where SM was barely detected. Furthermore, non-tumor cells contain higher SM levels than a wide range of tumor cell lines. Altogether, these studies show that SM content in cells in a non-proliferating state (i.e., quiescent, non-tumor, and differentiated cells) is higher than in cells in a proliferating state (activated, tumor and non-differentiated cells) indicating that SM levels may play a role in the control of cell cycle progression. In this context, it has been demonstrated that while SM synthesis plays an active role in cell adhesion, its degradation after SMase treatment looses attached cells, indicating that SM level is indeed important in cell adhesion regulation. The wide range of cell lines where this phenomenon occurs suggests that SM content regulation may be a general strategy used by cells to regulate their fate.

Finally, studies using acid sphingomyelinase (ASM-) KO mice demonstrate that SM levels are critical for brain functioning. Thus, the increased SM levels found in ASM-KO neurons results in reduced spine number and size and, low levels of filamentous actin. Overall, it is clear that the role of SM in cell function goes beyond being a mere structural element or a source of signaling molecules.

A. Introduction

Sphingomyelins (SM) are among the most abundant components of cellular membranes. In organisms ranging from mammals to protozoa, SM accounts for 10 to 20% of total cellular phospholipid content. It is preferentially concentrated in the outer leaflet of the plasma membrane providing so for a unique lateral asymmetric structure in the bilayers [1]. SM high packing density and affinity for sterols provide a rigid barrier to the extracellular environment and it is key in the formation of lipid rafts. SM consists of a sphingoid backbone [2-4], linked at the C2 position to a fatty acid via an amide linkage and to a phosphocholine moiety bound to the primary alcohol. Up to four different sphingoid bases have been described in human cells that combined with additional structural variations (hydroxylation, methyl-branching, acyl chain length) may result in 20-30 different SM-molecular species depending on the cell type [5]. Currently, it is accepted that in addition to its structural functions, SM serves as a source of important signaling molecules, basically as a substrate of the sphingomyelinase (SMase) which hydrolyzes SM into ceramide and phosphocholine. However, despite that evidence pointing to this relevant role in cell division has been available since the late 1980's, it has been clearly underestimated the possibility that SM by itself could be important in regulating crucial cell events. The rapid advances in mass spectrometry applied to lipidomic analysis provide comprehensive and detailed information regarding lipid composition of lipid extracts. Now it is also possible to assess the lipid composition of tissue sections thanks to imaging techniques based on mass spectrometry [6]. In any case, all results clearly show that lipid composition is highly tissue/cell dependent. However, little is known about the mechanisms regulating this specificity (Figure 1). This applies for all lipids including, obviously, SM. A recent study showing the direct interaction between a specific SM molecular species (SM d18:1-18:0) and a protein involved in COPI-transport (p24) [7] emphasizes the importance that the knowledge on the exact lipid composition and how this is achieved has. It is clear that the different molecular species of a particular lipid class may be as relevant as the different isoforms of a protein. Therefore, it is very important to fill the large gaps existing in our knowledge on how lipid composition is regulated.

B. Sphingomyelin Structural and Compositional Features

Despite of being one of the main components of the plasma membrane, not too much attention was paid to SM until 1988, when the concept of lipid rafts emerged [9]. The origin of the interest was some studies on epithelial cell polarity. The central idea was the existence of dynamic assemblies of cholesterol and sphingolipids at the outer membrane [10, 11]. Interestingly, SM not only interacts with cholesterol but sphingolipids are also important regulators of plasma membrane and cell cholesterol homeostasis [12, 13].

In two thorough reviews, Slotte describes precisely what are the structural characteristics that make of SM so suitable for its functions in cell membranes [5, 14]. Thus, it seems that the many roles played by SM in cell membranes are somehow related to the specific and unique combination of molecular properties for this sphingolipid. These properties include: a) low degree of unsaturation in both the long-chain base and the N-linked acyl chains, which translates to a high order parameter for the hydrophobic chains.

This high order, in turn, attracts cholesterol which helps to keep the chains in a liquid-ordered state; b) the specifically oriented hydrogen bonding sites provided by the 2NH and the 3OH of the long-chain base (D-erythro); c) the large phosphocholine head group of SM that appears to be optimal for SM/cholesterol interactions; d) the long-chain SM species have an inherent asymmetric nature because of the length mismatch between the sphingoid base and the usually long N-linked acyl chain in SMs. This interdigitation helps to explain how the inner leaflet is coupled to the outer leaflet. Importantly, modifications in this sense have been involved in cell division regulation. Kobayashi's group has been studying the distribution of SM across the plasma membrane for many years. They showed that during the cleavage furrow formation, SM of the outer leaflet might interfere with phosphatidylinositol bisphosphate (PIP_2) assemblies of the inner membrane. This interaction seems to be critical for cytokinesis [15].

From the compositional point of view, SM structure can be divided in two basic units: the sphingoid backbone (also referred as long-chain base) that is N-acylated with a fatty acid at the position C2. The most common long-chain bases are sphingosine (1,3-dihydroxy-2-amino-4-octadecene), followed by sphinganine (which lack the C4 trans double bond of sphingosine) and in lesser amount, phytosphingosine (having a hydroxyl group on C4) and 4,8-sphingadiene.

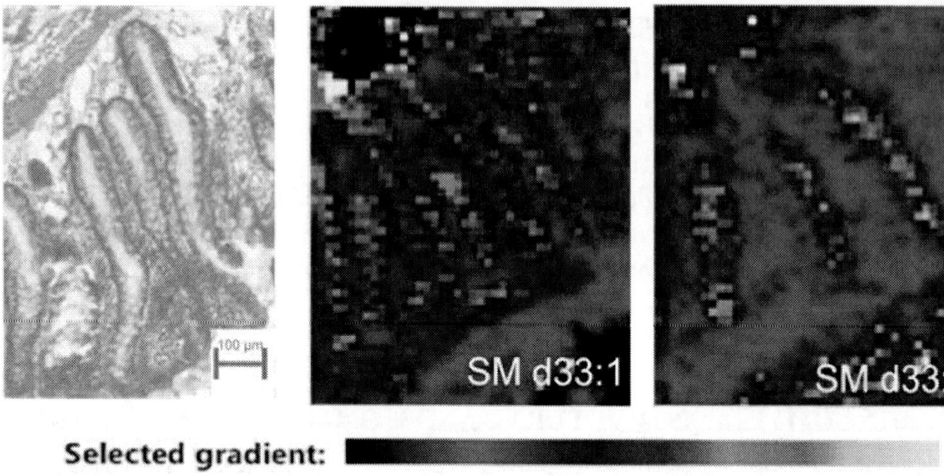

Figure 1. Specific distribution of SM molecular species in colon revealed by imaging techniques based on mass spectrometry. Comparison of the optical image of a consecutive section of a human colon tissue section stained with hematoxylin and eosin and the distribution of some selected lipid species, SM d33:1, m/z: 687.5 and SM d33:0, m/z 689.5, detected in negative ion mode by MALDI-imaging mass spectrometry (Matrix-assisted laser desorption/ionization, unpublished data). Note that structurally, the difference between species is based on the presence or absence of a single double bond and that this has a clear impact on the lipid distribution within the section. Scans were done at 10 µm pixel of spatial resolution. The scale bar is also included. All technical details may be found in Garate et al. [8].

The reason to find different sphingoid base has to do, as we will see later, with relative substrate promiscuity of the enzyme responsible for ceramide synthesis, the serine palmitoyl transferase (SPT). Thus, in addition to the canonical substrates, SPT is also able to metabolize other acyl CoAs in the range of C_{12} to C_{18}, resulting in a spectrum of sphingoid bases with variable carbon chain length [16, 17]. The acylation of dihydro-sphinganine (dhSph), the synthetic intermediate is then catalyzed by one of several possible ceramide synthases (CerS), each with different acyl chain preference [32].

Regarding the fatty acid composition, palmitic acid (C16:0), stearic (C18:0), C24:0 and C24:1 are the most common fatty acids found (Table 1). It is interesting that there is certain tissue specificity even in the distribution between palmitic and stearic acid: the former is more common in peripheral tissues while the latter is enriched in neural tissue [14]. One peculiarity of the sphingolipid fatty acid composition is the presence of 2-hydroxy fatty acids synthesized by fatty acid 2-hydroxylase (FA2H). The presence of these fatty acids has been known since the early 1900's [18]. They are particular enriched in myelin sheath although they have been also found in various non-neuronal mammalian tissues and tumors [18]. Table 1 summarizes the studies showing cells and tissue SM fatty acid composition.

C. Sphingomyelin Metabolism: What Do We Know?

Ceramide is the chemical backbone of all sphingolipids and it is metabolized into complex sphingolipids. Thus, ceramide can be further modified to yield ceramide-1-phosphate (Cer-1-P), SM, and simple or complex glycosphingolipids, including the various gangliosides [14]. For this reason, sphingolipid metabolism is usually described having ceramides as metabolic hub. SM and ceramide form the production cycle named "ceramide/SM cycle," acting mutually as a substrate and a product (Figure 2). These two pathways, the de novo and the SMase pathway, have been subject of intense study. However, we know now that ceramide accumulation may come from the catabolism of complex sphingolipids that are reused back to produce ceramides. This is the so-called recycling or salvage pathway. These three pathways do not act independently and in fact there is a high degree of interconnection between them, with intertwined chemical reactions, which complicates their study [27, 28].

Table 1. Cell and tissue SM fatty acid composition

		Fatty acid	Analytical Method	Reference
Tissue	Neural Tissue	Stearic acid (18:0)	HPLC-MS	Zhou 2012 [19]
	Guinea Pig Harderian glands	2-hydroxyacyl	GLC and GC-MS	Yasugi 1991 [20]
	Mature mammalian testes	Very long chain fatty acid	TLC and column chromatography	Brenton S. Robinson 1991 [21]
	Human umbilical cord	16:0 acyl chains sterified at the 3OH	GLC	Kramer 1996 [22]
	Mouse central nervous tissue	Enrichment of 2-hydroxy fatty acids	FAB-MS and GLC	Bosio 1996 [23]
	Mouse kidney	Enrichment of 2-hydroxy fatty acids	FAB-MS and GLC	Bosio 1996 [23]
Cells	Human skin Fibroblasts	d18:1/16:0, d18:1/24:1, d18:1/24:0	HPLC-MS/MS	Valsechi 2007 [24]
	Rat Cerebellar Granular cells	d18:1/18:0, d18:1/16:0	HPLC-MS/MS	Valsechi 2007 [24]
	Rat spermatozoa tails	18:0 and 16:0	TLC and GC	Oresti 2011 [25]
	Adult human spermatozoa	Very long chain fatty acid. Enrichment of 2-hydroxy fatty acids	TLC and column chromatography	Brenton S. Robinson 1991 Sandhoff 2010 [26]

ESI-MS: Electrospray ionization-mass spectrometry; FAB-MS: Fast atom bombardment-MS; GC-MS: Gas chromatography-MS GLC: Gas liquid chromatographic HPLC-MS Liquid Chromatography-MS; TLC: thin layer chromatography.

Sphingolipids that have been known to work as bioactive lipids are related to various cell functions. Particularly, ceramide has many roles in cell death, cell differentiation, cell cycle arrest, and autophagy [2, 3, 30, 31]. For example, in 1989, Okazaki et al., showed that ceramide was produced from SM through neutral SMase (N-SMase) during the differentiation of human leukemia HL-60 cells induced by vitamin D_3 [29]. On the other hand, SM is reported to form specific regions on the cellular plasma membrane named lipid microdomains, and it works as a regulator in diverse ligand/receptor signaling [4]. Numerous extracellular stimuli control amounts of SM and ceramide through the regulation of the ceramide/SM cycle although its function *in vivo* remains unclear.

C.1. The *De Novo* Pathway: First Ceramide, Then Sphingomyelin

The first step in SM synthesis occurs in the endoplasmic reticulum and it involves the synthesis of ceramide after the condensation of L-serine and palmitoyl coenzyme A (16:0-CoA).

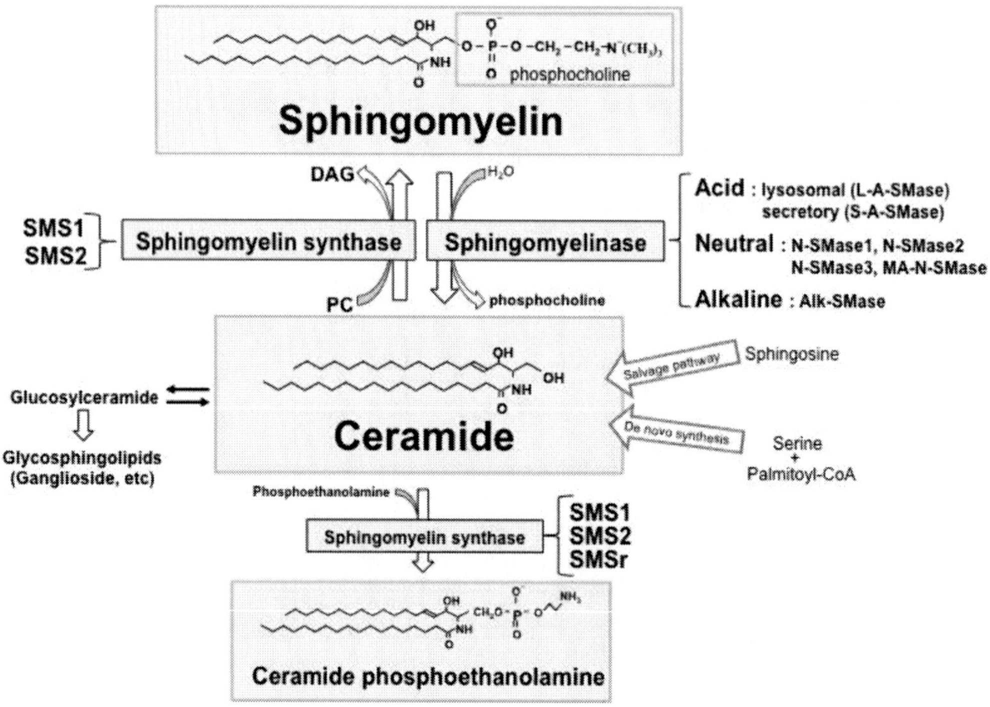

Figure 2. Ceramide/Sphingomyelin cycle. Ceramide is the center for sphingolipid metabolism to generate sphingomyelin (SM), ceramide phosphoethanolamine (CPE), and glycosphingolipids such as gangliosides (GM). There are three ceramide generating pathways; *De novo* synthesis from serine and palmitoyl-CoA, salvage pathway from sphingosine, and ceramide/SM cycle through SM hydrolysis by sphingomyelinase (SMase). Sphingomyelin synthase (SMS) has three isoforms; SMS1, SMS2, and SMSr. SMS1 and SMS2 generate SM and diacylglycerol (DAG) from ceramide and phosphatidylcholine (PC). All members of SMS have CPE synthesis activity. On the other hand, there are three SMase: Acid SMase (A-SMase), Neutral SMase (N-SMase), and Alkaline SMase (Alk-SMase). In turn, acid SMase has two isoforms: lysosomal form (L-A-SMase) and secretory form (S-A-SMase). Finally, there are four isoforms for N-SMase: N-SMase1, N-SMase2, N-SMase3, and mitochondrial N-SMase (MA-N-SMase).

This reaction is catalyzed by SPT generating 3-ketodihydrosphingosine, which is then reduced to dhSph by 3-ketosphinganine reductase. Under certain conditions, SPT can also use other amino acids, such as L-alanine or glycine, as alternative substrates [16]. This generates an atypical class of 1-deoxysphingolipids (1-deoxySL), which lack the C_1 hydroxyl group of canonical sphingolipids. 1-deoxySL are neurotoxic and they impair neurite formation in cultured neurons [16]. Elevated 1-deoxySL levels were found in plasma and lymphoblasts of hereditary sensory and autonomic neuropathy type I (HSAN1) patients and in plasma and peripheral nerve tissue of transgenic mice [32]. In addition, elevated 1-deoxySL levels were found in individuals with the metabolic syndrome and type-2 diabetes and they might be involved in the pathology of diabetic neuropathy as well [16, 33].

Then, the sphingoid backbone may be subsequently *N*-acylated by different ceramide synthases (CerS) producing (dihydro)-ceramides (dhCer), which are finally reduced to ceramide. Six different isoforms of CerS have been described so far, each of them displaying specificity toward fatty acyl-CoAs of defined chain length but not toward the hydroxylation and saturation status [34, 35]. It seems that CerS are less specific towards the long chain base than towards acyl-CoA [34, 35]. At this point, the dhCer undergoes desaturation by dhCer desaturase (the *cis* 4–5 double bond is inserted in mammals) to generate ceramides. All these reactions occur on the cytosolic surface of the ER. Therefore, CerS are responsible for the ceramide fatty acid composition.

Evidence is currently emerging that different CerS, and consequently ceramide containing specific fatty acids, play different roles in mediating specific biochemical events and thus they may have specific roles in the cell's physiology [36]. In one of the earliest studies to examine cell response to ceramide molecular species, it was observed that B-cell receptor triggering in lymphocytes induced a biphasic elevation in ceramides [37]. The early phase -minutes/hours- consists primarily of C16-ceramide *de novo* synthesis and preceded the onset of activation-induced cell death. The late phase involves C24-ceramide accumulation after cell death activation. Among all ceramide species, only C18-ceramide is selectively downregulated in head and neck cancer tissues [38]. Recently, CerS5 and CerS6 have been implicated in mediating cytotoxic responses to ionizing radiation whereas CerS2, probably acting in mitochondria, offered partial protection [39]. Similarly, CerS5 and CerS6 are also involved in ceramide generation by the salvage pathway being necessary for regulation of the membrane permeability in the programmed cell death execution phase in MCF-7 breast cancer cells in response to UV radiation [40]. In addition, the first study to demonstrate biological activities for dhCer shows that C2-dhCer treatment induces autophagy in prostate cancer cells [41]. In summary, these studies not only involve each one of the CerS isoenzymes in the formation of specific ceramides, but they also suggest specific functions for each ceramide molecular species.

Once synthesized in the endoplasmic reticulum (ER), ceramides are delivered to the Golgi apparatus where they are converted by SM synthase (SMS) or by GlcCer synthase to SM or to glucosylceramide (GlcCer), respectively. Ceramides must travel from the ER to the Golgi apparatus by facilitated mechanisms because their hydrophobic nature prevents the spontaneous transfer through the cytosol [42]. The transport of ceramides to the Golgi complex occurs either through vesicular transport, which delivers ceramide for the synthesis of GlcCer, or through the action of the transfer protein CERT, which specifically delivers ceramides for SM synthesis [43]. CERT is able to specifically extract ceramide from the ER and it contains a pleckstrin homology (PH) domain allowing its targeting to the Golgi [30, 44, 45]. GlcCer synthase catalyzes the transfer of glucose from UDP-glucose to ceramide. This enzyme resides in the *cis* Golgi and its active site is oriented toward the cytosol [46]. After translocation to the Golgi lumen, GlcCer is converted to more complex glycosphingolipids. The Golgi-associated four-phosphate adaptor protein 2 (FAPP2) was identified as an essential component of the glycosphingolipid synthetic machinery, as it mediates non-vesicular transport of GlcCer from its site of synthesis at the early Golgi to distal Golgi compartments, where most of the enzymes involved in complex GSLs assembly are located [47]. Finally, SM production involves the enzymatic transfer of a phosphocholine head group from phosphatidylcholine to the primary hydroxyl of ceramide, yielding diacylglyeron in the process (Figure 2, [48]). The enzyme catalyzing this reaction, phosphatidylcholine: ceramide choline-phosphotransferase, or SMS, occupies a central position in sphingolipid and glycerophospholipid metabolism. Since the enzyme is also able to catalyze the reverse reaction, the formation of phosphatidylcholine and ceramide from SM and diacylglycerol [49], SMS has considerable biological potential as regulator of the pro-apoptotic factor ceramide and the mitogenic factor diacylglyerol [48, 50].

In 2004, SMS (Sgms) genes were cloned for the first time, and three isoforms were confirmed: SMS1 (Sgms1), SMS2 (Sgms2), and SMS-related (SMSr, Samd8) (Table 2) [48, 51]. Two SMS isoforms with close structural similarities and a 57% sequence homology have been described to date, SMS1 and SMS2. SMS1 is mostly associated with the Golgi apparatus, and SMS2 is mainly distributed to the plasma membrane and to a lesser extent to the perinuclear region [48]. Both enzymes have the active site facing the exoplasmic leaflet of the membranes. While the Golgi enzyme would account for generating the bulk of cellular SM, the second enzyme may serve a principal role in signal transduction at the plasma membrane. Remarkably, a difference in substrate preference between both isozymes has been described.

Thus, SMS1 uses phosphatidylcholine as a substrate, whereas SMS2 can use both phosphatidylcholine and phosphatidylethanolamine [52]. SMS1 and SMS2 are encoded by ubiquitously expressed genes and they are uniformly distributed among mammalian tissues [48]. In vitro, SM generated by SMS1 or SMS2 is reported to play a role in regulation of cell proliferation, death and migration through lipid microdomain modification [53, 54]. However, its pathophysiological meaning and the mechanism of action remain to be clarified.

In addition to SMS1 and SMS2, there is a third SMS-related (SMSr) protein. SMSr has no SMS activity but it catalyzes ceramide phosphoethanolamine (CPE) synthesis from ceramide and phosphatidylethanolamine in the ER [55]. The activity of CPE synthesis is faintly observed in both SMS1 and SMS2 [8, 9]. In addition, since the amount of cellular CPE is quite low, the production of CPE seems to play a role in ceramide homeostasis rather than in generation of bioactive lipids. Recently, Ding et al. reported that SMS1 also has CPE synthase activity as same as SMSr, and SMS1-KO showed a decrease of plasma CPE [56]. At cell levels, SMSr has a role for maintenance of the ceramide homeostasis and cell survival through ceramide recognition and CPE production. The biological functions of CPE and its synthesis *in vivo* are also under elucidation.

C.2. SMase Pathway

Once synthesized, SM and complex glycosphingolipids are transported to the plasma membrane via vesicular trafficking. There, SM can be metabolized back to ceramide either by acidic sphingomyelinase (A-SMase), on the outer leaflet of the membrane, or by N-SMase, which resides on the inner leaflet of the bilayer [57]. There are three types of SMase, classified according to the differences in enzymatic optimal condition: acid-, neutral and alkalin-SMase (A-, N-, Alk-SMase) (Table 3).

Table 2. SMSs main features

		SMS1	SMS2	SMSr
Gene symbol (Locus)		*Sgms1* (Chr.19 C1)	*Sgms2* (Chr.3 G3)	*Samd8* (Chr.14 B)
Amino Acids		419	365	432
Cellular localization		Golgi	Golgi, Plasma membrane (PM)	ER
Expression (high)		Ubiquitous (islet, testis)	Ubiquitous (liver, kidney, intestine)	Ubiquitous (brain)
Enzymatic reaction	Substrate ↓ Product	Ceramide + PC ↓ SM + DAG; Ceramide + PE ↓ CPE	Ceramide + PC ↓ SM + DAG; Ceramide + PE ↓ CPE	Ceramide + PE ↓ CPE

CPE, Ceramide Phosphoethanolamine; DAG, Diacylglycerol; ER, Endoplasmic reticulum; PC, Phosphatidylcholine; PE, Phosphatidylethanolamine.

Table 3. SMases main features

	Acid		Neutral				Alkaline
Protein	L-A-SMase (lsysosme)	S-A-SMase (secreted)	N-SMase1	N-SMase2	N-SMase3	MA-N-SMase	Alk-SMase
Gene symbol (Locus)	*Smpd1* (Chr.7 E3)		*Smpd2* (Chr.10 B2)	*Smpd3* (Chr.8 D2)	*Smpd4* (Chr.16 B1)	*Smpd5* (Chr.15 D3)	*Enpp7* (Chr.11 E2)
Amino acids	627		419	655	823	483	446
Cellular localization	Lysosome /Endosome PM	Extracellular	ER	Golgi PM	ER Golgi PM	Mitochondria	unknown
Expression (High)	Ubiquitous		Ubiquitous (kidney)	Ubiquitous (brain, spleen)	Ubiquitous (muscle, heart)	Ubiquitous (testis, pancreas, brain)	Intestinal mucosa
Enzyme Substrate/product	SM/Ceramide		SM/Ceramide				SM/Ceramide
Cation	no need Zn^{2+}	Zn^{2+}	Mg^{2+}	Mg^{2+}	Mg^{2+}	Mg^{2+} • Mn^{2+}	Inhibition by Zn^{2+}
pH	pH 5.0		pH 7.5	pH 7.5	pH 7.5	pH 7.5	pH 9.0

Currently, seven distinct SMase have been identified based on their optimal pH for activity, subcellular localization, and cation-dependence. A-SMase (Smpd1) and Alk-SMase (Enpp7) have one isoform respectively while N-SMase has four isoforms: N-SMase1 (Smpd2), N-SMase2 (Smpd3), N-SMase3 (Smpd4), and mitochondria localized MA-N-SMase (Smpd5) [59]. A-SMase, widely investigated as the responsible enzyme of Niemann-Pick Disease (NPD) type A and type B, was first described in 1967 [58] and cloned in 1989 [60]. Mammalian N-SMase1 was cloned in 1998, and four isoforms in N-SMase were described later: N-SMase1 [61], N-SMase2 [62], N-SMase3 [63], and mitochondrial MA-N-SMase [64]. In 1969, Nilsson et al. found Alk-SMase as the enzyme that digests dietary SM into ceramide in small intestinal mucosa [65] and cloned in 2005 [66].

Ceramide generation by SMase is enhanced by various stimuli such as anti-cancer drugs, oxidative stress and cytokine stimulation, regulating cell growth, death, differentiation, inflammation and autophagy [4].

As it was previously mentioned, ceramides formed during both the *de novo* synthesis and the salvage pathway are then transformed in a long list of metabolic intermediates, some of them with high bioactive potential. Thus, ceramide-1-P (Cer-1-P) is implicated in cell differentiation [67, 68], apoptosis and autophagy [69-73], whereas sphingosine (Sph) has been proposed to induce cell death by a ceramide-independent mechanism [74]. Both ceramide and Sph can be reversibly converted to Cer-1-P by ceramide kinase and to sphingosine-1-phosphate (Sph-1-P) by sphingosine kinase, respectively [75]. Sph-1-P has opposite effects promoting cell proliferation, survival, and inhibition of apoptosis. Sph-1-P acts as a ligand for a family of G protein-coupled receptors, as well as a second messenger [76, 77]. Cer-1-P is a potent inhibitor of apoptosis and it induces cell survival by activating $cPLA_2$ and inhibiting aSMase in a cell-dependent manner [78, 79]. Importantly, the improvement in analytical techniques is allowing the quantification of compounds that before were not taken into account because of their low levels but that they seem to be playing important roles such as dihydrosphingomyelin (dhSM) or dihydroceramide (dhCer). For example, an increase in dhSM inhibits HIV-1 cell infection, because it gives rise to more rigid membranes that inhibits viral-cell membrane fusion [80].

C.3. Salvage Pathway

From plasma membrane, sphingolipids may be recirculated through the endosomal pathway, via the recycling or salvage pathway [28]. Thereby, SM and GlcCer are metabolized back into ceramide in the lysosomal compartment by aSMase and glucosidases. Ceramide is then degraded by acid Ceramidase (CDase) to form Sph [44, 46]. Due to its ionizable positive charge, the salvaged Sph is able to leave the lysosome and shows adequate solubility in the cytosol to move between membranes, including the ER where it would be available for recycling [28]. These bases serve also as substrates for sphingosine kinases that form sphingosine-1-phosphate (Sph-1-P).

C.4. Nuclear Sphingolipid Metabolism

Many aspects of the glycerophospholipid metabolism and function in the cytoplasm are recapitulated in the nucleus [81]. The nuclear envelope contains the large bulk of nuclear lipids, which, in addition to providing structural support and isolation of nuclear functions, are also the source of numerous signaling reactions. Nuclear SMS and SMase activities have been described and according to the data, it seems that nuclear and cell SM metabolism are oppositely regulated. Thus, in the rest of the cell, SM degradation leads to apoptosis induction, mostly through the ceramide accumulation in mitochondria. SM, the most abundant lipid in chromatin with respect to nucleosomes and nuclear membranes, is localized near to newly synthesized RNA, and it appears to protect it from RNase digestion. In rat hepatocytes, SM synthesis increases during the differentiation process and apoptosis, while an increase of SMase activity, with a concomitant decrease in SM, is observed at the beginning of S-phase of the cell cycle, indicating a possible role in stabilizing the DNA double helix [82]. However, recent studies of the same group suggest that the enhancement in DNA synthesis is primarily due to the formation of SM-enriched nuclear membrane microdomains, which act as membranous platforms for gene expression in a similar way as lipid rafts.

C.5. Regulation of Sphingolipid Metabolism

The fact that sphingolipids are implicated in such important processes for the cell, points towards the need of a fine regulation mechanism. In fact, the balance between the relative levels of the bioactive sphingolipids, i.e. ceramide, Sph, Sph-1-P and Cer-1-P, has been proposed to act as a 'rheostat' to determine the cell fate with respect to survival [41, 83]. Orm proteins were identified as essential components of the sphingolipid rheostat that allows cells to fine-tune sphingolipid synthesis according to their needs. Thus, Orm proteins form a conserved complex with SPT that is regulated by a phosphorylation cycle of Orm proteins relieving their inhibitory activity when sphingolipid production is disrupted [84].

The sphingolipid metabolism rheostat can be also deregulated by defects in enzymes and proteins required for lysosomal catabolism that lead to inherited diseases, the lysosomal storage diseases. Indeed, human diseases caused by alterations in the metabolism of sphingolipids or glycosphingolipids are mainly disorders of the degradation of these compounds. Lysosomal storage diseases can be classified according to the stored substances, as sphingolipidoses, mucopolysaccharidoses, mucolipidoses, glycoprotein and glycogen storage diseases. However, in most lysosomal diseases the stored material can be rather heterogeneous [85-87]. As a result of the growing amount of accumulating material, a variety of pathogenic cascades are activated such as altered calcium homeostasis, oxidative stress, inflammation, altered lipid trafficking, autophagy, endoplasmic reticulum stress, autoimmune responses subsequently leading to cell death [88].

D. Evidences Pointing to the Relevance of SM in Cell Physiology

Because of its presence in plasma membrane microdomains SM metabolism has been involved, directly or indirectly, in so many cell processes that it is not possible to cover all of them in a single revision. For this reason we decided to choose two aspects, cell division and pathogen invasion, which we considered will help us transmitting the importance that this sphingolipid in particular has in cell physiology.

D.1. SM and Cell Division: Proliferation, Differentiation and Tumorigenesis

Evidence that SM was somehow involved in the control of cell division began to appear back in the late 80's early 90's. Thus, Dressler et al. showed a selective increase in SM levels within the adherent macrophage population during phorbol ester (TPA)-induced differentiation. Interestingly, the effect of phorbol esters on these sphingolipids was biphasic: an initial burst phase of SM synthesis correlated with rapid reduction in ceramide levels that was followed by a prolonged phase of SM synthesis correlated with additional ceramide synthesis. The lower levels of ceramide and phosphatidylcholine concurrent with SM synthesis suggest that, the so-called at that time, phosphatidylcholine: ceramide choline phosphotransferase activity was enhanced by treatment with phorbol ester at early time intervals [89]. Similarly, Qureshi et al. showed that the differentiation of HL-60 induced by DMSO and retinoic acid increases SMS activity [90]. Prior studies from Okazaki et al. showed that vitamin D_3-induced differentiation of HL-60 cells into monocyte/macrophages also involved SM metabolism [29]. However, contrary to the effects of phorbol esters, vitamin D_3 action appeared to involve SM degradation via a N-SMase [29]. In 1992, Olivera et al. showed that both treatment with exogenous SM as well as with ceramide cell-permeable analogs produce a mitogenic effect in quiescent Swiss 3T3 fibroblasts and it potentiates the action of several growth factors [91]. On the other hand, quiescent satellite cells (skeletal muscle stem cells) present high SM levels in the plasma membrane that fall once cells are activated and proliferate [92].

Table 5. Changes on SM molecular species during pathology

Disease	SM features	Method	Reference
Human Lung Cancer	36:1, 42:2, 42:1, 40:1, 36:2 and 34:1	ESI-MS 2D-imagingMS	Erya Marien 2015 [95]
Human Colon Cancer	Decrease in SM	^{31}P MRS	Merchant 1991 [96]
Human Breast Cancer	Decrease in SM	^{31}P MRS	Merchant 1991 [96]
Esophageal Tumor	Decrease in SM	^{31}P MRS	Merchant 1993 [94]
Alzheimer disease	Decrease in SM	ESI-MS	Gutler 2003 [97]
Mouse obesity	Decrease in adipose SM	HPLC/TMS	Samad 2006 [98]
Lung emphysema	Increase of ceramide		Petrache 2005 [99]
Human lung adenocarcinoma cells	Low SM content	TLC	Barceló-Coblijn 2011 [93]
Human lymphoblastic leukemia cells	Low SM content	TLC	Barceló-Coblijn 2011 [93]
Human glioma cells	Low SM content	TLC	Barceló-Coblijn 2011 [93]
Breast cancer cells	Decrease in d18:1/20:4, d18:1/22:0 and d18:1/24:1 in more invasive cells	ESI-MS	Dória 2013 [100]
MC4-L2 and MC4-L5 breast cancer cells	Decrease in d18:1/22:0, d18:1/24:1, d18:1/24:0	TLC and ESI-MS	Dória 2011 [100]
PC3 prostate cancer cells	Low SM content Most abundant species: d18:1/16:0, d18:1/24:1, d18:1/24:0	UHPLC-MS	Llorente 2013 [101]
Acute myeloblastic leukemia cells	Increase of 18:2, 18:3 and 20:4	GC	Klock 1979 [102]

ESI-MS: Electrospray ionization-mass spectrometry; FAB-MS: Fast atom bombardment-MS; GC-MS: Gas chromatography-MS GLC: Gas liquid chromatographic HPLC-MS Liquid Chromatography-MS; MRS: Magnetic resonance spectrometry; TLC: thin layer chromatography.

Finally, a wide range of tumor cell lines have shown to contain higher SM levels than non-tumor cells [93]. However, caution should be taken in this sense because a thorough revision of the literature showed up some contradictory results, suggesting that SM levels and its alteration associated to pathological processes most probably are cell- or tissue-dependent

(Table 5). Altogether, these studies show that SM content in cells in a non-proliferating state (i.e., quiescent, non-tumor, and differentiated cells) is higher than in cells in a proliferating state (activated, tumor and non-differentiated cells) indicating that SM levels may play a role in the control of cell cycle progression. Unfortunately most of the analyses were done using techniques that did not allow to distinguish if the differences in SM content were due to a general effect on the SM lipid class or specific to a particular SM molecular species [94].

There are two moments during cell division when SM seems to play a prominent role: cytokinesis and cell (de)attachment.

Cytokinesis: Furrow Cleavage Regulation

As a cell divides, it undergoes massive shape changes, culminating in the formation of a small structure, the midbody, where cleavage between daughter cells occurs [103]. Cytokinesis is the final act of cell division whereby the cytoplasm is pinched apart by constriction of an actin based contractile ring [104]. Although membranes are intimately connected to the cytoskeleton and intact membranes are an absolute requirement after division has been completed, little is known about the role of membranes during cell division. The contractile ring is intimately associated with the plasma membrane to which its assembly and contraction must be anchored and which is the site of new membrane addition during late cytokinesis [105, 106]. Cytokinesis has to achieve two crucial processes: equal sorting of cellular components between the daughter cells, and membrane ingression and closure [103]. Both processes strongly rely on membrane components and properties [104, 107]. Although it is known that cell membranes undergo dramatic structural rearrangements during cytokinesis, and it is obvious that membrane rearrangements are needed to seal daughter cells after severing, little is known about whether (and how) specific lipids are involved in cytokinesis.

Work by Umeda's and Kobayashi's groups using fluorescent compounds that bind to specific membrane lipids show that localization of different membrane lipid components is modified during cell division. Thus, phosphatidylethanolamine [108], cholesterol, SM and PIP_2 [15] accumulate in the midbody during cytokinesis. According to their results, there is a colocalization between the SM-rich domains in the outer leaflet and PIP_2-enrich domains in the inner leaflet. The latter is required for the recruitment of factors needed for cytokinesis to proceed as RhoA [109, 110]. Using model membranes, Makino et al. (2015) described that two SM-binding toxins, lysenin (Lys) and equinatoxin II (EqtII) differentially bind to SM depending if SM is clustered or dispersed, respectively [111]. Using derivatives of these probes, the authors demonstrated that the accumulation in the midbody of clustered, but not dispersed, SM is needed for cytokinesis to proceed. Atilla-Gokcumen and coworkers used a completely different approach to assess the changes occurring in cell division [112]. They analyzed by liquid chromatography-mass spectrometry the lipid composition not only at the different phases of the cell cycle but also from isolated midbodies. Results showed different lipid species that were accumulated in the midbodies C22- and C24-ceramide, C16- and C24-hexosylceramide, phosphatidic acid (16:0/16:0), triacylglycerol (16:1, 12:0, 18:1), phosphatidylserine (18:0/20:4). The meaning of these differences is still unknown.

Certainly, the development of future imaging techniques able to analyze individual cells at subcellular resolution will help us to understand better how and why lipid composition changes during cell division [93, 113].

SM in the Regulation of Cell Adhesion Proteins Location

The regulation of cell adhesion is critical for an adequate cell division and the loss of it leads to uncontrolled cell division, i.e., to a tumor formation. In this context, while SM synthesis plays an active role in cell adhesion [89], its degradation after SMase treatment loosed attached cells [91], indicating that SM levels may play an important role in cell adhesion. The wide range of cell lines where this phenomenon has been demonstrated suggests that the regulation of SM content may be a general strategy used by cells to regulate its fate. Next we summarize the results obtained regarding the implication of SM in the regulation of three well know systems responsible for cell adhesion: integrins, focal adhesion and adherens junctions.

Integrins are transmembrane receptors that mediate cell-cell adhesion, cell attachment to the extracellular matrix, cell spreading and cytoskeletal organization. Integrin-mediated adhesion initiates transduction of signals towards the cell interior and also regulates transmission of signals from growth factor receptors [114]. Thus, activation of most signaling pathways including the Erk, PI 3-kinase, FAK, Src and Rho GTPases pathways depend on integrins [115]. In turn, over the last few years it has been demonstrated that cyclins and Cdks, which control specific checkpoints of the cell cycle, are not regulated only by receptor tyrosine kinases (RTK) but also by integrin signaling [116]. Interestingly, function of the Ras/Erk pathway, JNK, PI-3-kinase, FAK and Src family kinases, which are all highly dependent on integrins, have been linked to lipid rafts [117]. In fact, functional activation of integrins, that is, their conversion to the high affinity state, appears to be linked to lipid raft localization [117]. Consequently, it was proposed that integrins regulate lipid rafts and that this regulation may be important for the spatial control of cell migration and signaling pathways involved in anchorage-dependent cell growth [118, 119]. However, little is known about the effect of changes in lipid raft properties, in particular those associated to alterations in SM content, on integrin-regulated pathways.

There are different types of cell-matrix adhesion complexes mediated by integrin that differ in their basic features (as size, location, and shape), in their composition, dynamics and their linkages to F-actin. One of the best characterized is the focal adhesion (FA), which occurs by the binding of the integrin-extracellular domain to the extracellular matrix proteins followed by the interaction of β-integrin cytoplasmic domain with talin, which in turn, binds vinculin. Interestingly, FA complexes have been located in lipid rafts, which shows a clear enrichment in cholesterol and SM compared to the detergent soluble fraction (4-fold and 70% respectively) [120]. However, because most of the studies were focused on the effect of the alteration of cholesterol on FA stability, the role of SM in the stability of this protein macrocomplex has not been established.

Finally, in epithelial tissues, the "core complex" that organizes the structures responsible for the cell-to-cell adhesion is the adherens junction (AJ). AJ is a complex system of proteins including cadherins, α-, β-, and γ-catenins, α-actinin, and vinculin [121]. E-cadherin, which spans the plasma membrane, contacts E-cadherin on other cells and it connects with the actin cytoskeleton inside the cell. Different components of AJ are found membrane microdomains and changes in lipid composition clearly alter the cell-cell adhesion and consequently the epithelial integrity [122]. Further studies, using MDCK (Madin-Darby canine kidney) cells, demonstrated that the SM synthesis is in fact required for the acquisition of mature AJ structures. If affected, AJ are not able anymore to assemble and cells progress to the acquisition of a less mature phenotype [27].

D.2. SM in the Response to Pathogen Infections

The use of the host cells mechanism by pathogens has been known for a long time ago. As we will see, a common feature in the strategy used by pathogens to infect the host is to take advantage of its endocytic-vesicle trafficking system. The latter regulates the vesicular trafficking from Golgi apparatus to the plasma membrane and it is well known that is regulated by SM metabolism. Importantly, vesicle formation during immune/immunological cells activation is also critical.

In this context, it is well known that ceramides induce the formation of membrane invaginations responsible of the particle engulfment by phagocytic cells. For this reason most of the studies focus on ceramides relegating the role of SM to a mere ceramide precursor. However, evidence points out that both ceramide and SM are needed in the response to pathogens. Thus, membrane damages and the binding of microbial pathogens can result in the translocation of lysosomal A-SMase to the plasma membrane outer leaflet where it hydrolyzes surface-exposed SM [32]. The consequent increase in ceramides affects the physical properties of lipid rafts and other membrane-regulated processes as membrane invagination, receptor clustering and macropinosome formation [32], all of which promote the uptake of particles, including viruses, into cells.

There is a series of viruses that uses macropinocytosis for host entry including *Neisseria gonorrhoeae* [123], measles virus, and ebola virus. Interestingly, in addition to Rhinoviruses, they all seem to share an A-SMase-dependent cellular entry mechanism [32, 124]. Consistently, the inhibition of lysosomal exocytosis prevents ebola virus infection. In 2010, Qureshi et al. investigated the role of host sphingolipids during an infection caused by the fungal pathogen *Cryptococcus neoformans* in mice lung [90]. Their results show that SMS regulates the killing ability of neutrophils by controlling vesicle formation. Thus, SMS inhibition prevents the release of antifungal factors. Interestingly, the analysis of lung tissue sections by MALDI-imaging techniques showed that a specific SM molecular species was associated with the neutrophil infiltration (SM d18:1/16:0) while others were totally absent (SM d18:1/24:1).

In the same line, Tafesse et al. prove that the spreading of influenza virus requires a functional SM synthesis metabolism. Particularly, the latter is needed to drive influenza viral glycoproteins into specific lipid domains at the plasma membrane where the virus budding and release occurs [125]. In 2011, Fernandes et al. extend the list of pathogens affecting SM metabolism to protozoa [126]. *Trypanosoma cruzi*, an intracellular protozoan, invades cells by forming membrane-bound intracellular vacuoles independently of host cell actin rearrangements [127]. Upon host cell contact, *T. cruzi* triggers cytosolic calcium transients that induce exocytosis of lysosomes, containing among other enzymes A-SMase, mimicking so one of the processes of plasma membrane injury and repair [128]. Similar to what happens during some viruses infection, once at the outer leaflet A-SMase, alters the SM and ceramide content and consequently the physicochemical properties of the plasma membrane so that the pathogen entry is favored [126]. Finally, not only pathogens use the SM metabolism to inhibit the immune response. Recently, Okoye et al. demonstrate that the Foxp3^{+} T cells (an inhibitory T cell) use miRNa exosomes delivered to inhibit the immune response [129]. The exosomes secretion had been clearly related to the SM metabolism [101].

E. How Can SM Be Involved? Lessons from KO-Models

E.4. Sphingomyelin Synthases KO-Models

The production of animal disease models using the development of gene modification such as transgenic, knock-in, and knockout (KO) mice has allowed the analysis of physiological functions of sphingolipids in vivo. Many papers have been published reporting SMS1-KO and SMS2-KO mice and their phenotypes. Recently, SMSr-KO mice have been established and the physiological roles of SMSr and CPE activity in mice have been shown [143]. Table 6 and 7 summarize the phenotypes of mice genetically modifying SMS and SMase genes, and the effect in each mouse disease model on sphingolipids levels.

Sphingomyelin Synthase 1

SMS1 deficiency in mice showed moderate neonatal lethality (approximately 40% at 100 days after birth), reduction of body weight, and decrease of white adipose tissue (WAT) [130, 131]. In terms of sphingolipid metabolism, SMS1-KO reduced SM content and conversely increased ganglioside GM3 in pancreatic islets and WAT. In addition, SMS1-KO also exhibited a decrease in SM levels in plasma, macrophages and liver compared with those of wild type mice (59%, 45%, and 54%, respectively) [132]. Inversely, GlcCer and GM3 levels were increased by 4- to 12-fold. In agreement with these observations, Okazaki's group confirmed a SM reduction and a rise in glycosphingolipids in liver, brain, and blood cells such as platelets of SMS1-KO mice (unpublished data). In 2011, Yano et al. published the first report of the phenotypes in SMS1-KO mice [130]. Deficiency of SMS1 induced mitochondrial dysfunction by the accumulation of reactive oxygen species (ROS) and following inhibition of insulin secretion in pancreatic β cells. Treatment of SMS1-KO with N-acetyl cysteine, a well-known anti-oxidant reagent, blocks ROS production as well as the inhibition of insulin release. It is thought that SMS1 acts as a regulator of ROS generation but has no direct cytotoxicity. ROS induced by SMS1 deficiency caused the reduction of insulin secretion, resulting in metabolic deficiency via decrease of glucose uptake in the whole body. SMS1-KO revealed ROS production and mitochondria dysfunction in WAT, supporting the SMS1 function as ROS regulator [131]. These phenomena are thought to be related to the decreases in body weight and a moderate neonatal lethality.

Li et al. found a decrease in SM content in macrophages of SMS1-KO mice and they focused their research on the inflammatory response in these cells [132]. Lack of SMS1 in macrophages decreased the release of cytokines such as interleukin 6 (IL-6) and tumor necrosis factor-α (TNF-α) in response to lipopolysaccharide (LPS). In addition, bone marrow transplantation from SMS1-KO to low density lipoprotein receptor-KO mice ($Ldlr^{-/-}$), which is a mice model of atherosclerosis, palliated the exacerbation of atherosclerosis in aorta, suggesting the implication of SMS1 in cytokine release and in the maintenance of bone marrow-derived inflammatory cells such as monocytes or macrophages. On the other hand, Dong et al. showed that SMS1 deficiency inhibited Concanavalin A (ConA)-mediated hepatitis by the suppression of the activated CD4-positive T cells [133]. SMS1-KO decreased SM level in lipid microdomains and inhibited the aggregation of T cell activation complex such as, linker for activation of T cells (LAT) and T cell receptor (TCR). Thus, the suppression of cytokine production in T cells, such as IL-2 and interferon-γ (IFN- γ), was suggested to be the cause of lower inflammatory response in SMS1-KO mice. SMS1-regulated SM levels of macrophages lipid microdomains might be controlling inflammation, as microdomains are an essential site for generation of the inflammatosome signaling.

Table 6. Summary of the phenotypical characteristics described in SMS – KO models

	Phenotype	References
SMS1	Moderate neonatal lethality, ↓ Body weight, ↓ White adipose tissue (WAT) ↓ Insulin secretion from pancreas ROS accumulation and mitochondrial dysfunction in β-cells	Yano et al. *J.Biol.Chem.* 2011 [130]
	↓ SM in plasma, liver, macrophage, ↑ GlcCer & GM3 ↓ Atherosclerosis in *Sms1*$^{-/-}$ → *Ldlr*$^{-/-}$ mice	Li et al. *Arterioscler.Thromb.Vasc.Biol.*2012 [132]
	Hearing impairment Atrophy of stria vascularis, Disorganization of marginal cells	Lu et al. *J.Physiol.* 2012 [134]
	Dysfunction of CD4+ T-cell, ↓ Hepatitis induced by ConA	Dong et al. *Int. Immunol. 2012* [133]
	ROS accumulation and mitochondrial dysfunction in WAT	Yano et al. *PLOS one* 2013 [131]
	↓ Clot retraction in platelets	Kasahara et al. *Blood.* 2013 [135]
SMS2	Attenuation of LPS-mediated NF-kB activation in macrophages	Hailemariam et al. *Arterioscler.Thromb.Vasc.Biol.*2008 [137]
	↓ Plasma SM ↓ Atherosclerosis in *Sms2*$^{-/-}$ → *Ldlr*$^{-/-}$ mice	Liu et al. *Circulation Res.*2009 [139]
	↓ Atherosclerosis in *Sms2*$^{-/-}$;ApoE$^{-/-}$ mice	Fan et al. *Arterioscler.Thromb.Vasc.Biol.*2010 [138]
	Attenuation of LPS-induced lung injury	Gowda et al. *Am.J.Physiol.Lung.Cell. Mol.Physiol.* 2011 [136]
	↓ SM in brain, ↓ P-glycoprotein (Mdr-1)	Zhang et al. *Biochem.Pharmacol.* 2011 [140]
	↓ Liver steatosis and type II diabetes induced by HFD	Mitsutake et al. *J.Biol.Chem.* 2011 [141]
	↓ SM and ↑ Ceramide in liver, adipose tissue, skeletal muscle ↑ Insulin sensitivity, ↓ Obesity induced by HFD	Li et al. *Mol. Cell. Biol.* 2011 [142]
	↓ Liver steatosis induced by HFD, ↓ PPARγ2, ↓ FFA uptaking	Li et al. *Arterioscler.Thromb.Vasc. Biol.*2013 [188]
SMSr	**No change** in plasma CPE (↓ plasma CPE in SMS1-KO or SMS2-KO)	Ding et al. *J. Lipid Res.* 2015 [56]

Table 7. Summary of the phenotypical characteristics described in SMase – KO models

		Phenotype	References
Acid	**A-SMase**	↑ SM in liver & brain, ↑ Cholesterol in blood Niemann-Pick type A and B	Horinouchi et al. *Nat.Genet.* 1995 [146]
		Resistance of radiation-induced apoptosis	Santana et al. *Cell.* 1996 [150]
		↑ Liver metastasis of SL4 colon cancer	Osawa et al. *J.Clin.Invest.* 2013 [151]
Neutral	**N-SMase1**	**No change** in SM and ceramide	Zumbansen et al. *Mol.Cell Biol.* 2002 [153]
	N-SMase2	Dwarfism, delay of puberty by decrease of pituitary hormone **No change** in brain SM and ceramide Chondrodysplasia **Loss** of N-SMase activity in liver and bone	Stoffel et al. *Proc. Natl. Acad. Sci. USA.* 2005 [155]
		Osteogenesis & dentinogenesis imperfecta (Deletion of 33 amino acids in C-terminal) **(N-SMase1/2)**	Aubin et al. *Nat. Genet.* 2005 [156]
		Chondrodysplasia and Dwarfism	Stoffel et al. *Am. J. Pathol.* 2007[158]
	N-SMase3	None	
	MA-N-SMase	None	
Alkaline	**Alk-SMase**	↓ Digestion of dietary SM Hypertrophy of intestinal epithelium	Zhang et al. *J. Lipid Res.* 2011 [162]
		↓ Absorption of dietary SM and dietary cholesterol	Zhang et al. *Am. J. Physiol. Gastrointest. Liver Physiol.* 2014 [163]
		↑ Colon carcinogenesis in AOM/DSS model	Chen et al. *Mol. Cancer Ther.* 2014 [166]

Lu et al. reported that SMS1-KO exhibited hearing impairment caused by atrophy of the *cochlear stria vascularis* and the disorganization of marginal cells [134]. Hearing function is maintained by keeping positive charges, named endocochlear potential (EP), at high electric potential (+80 mV) in inner lymph fluid filling the inner ear cochlea. The maintenance of EP is related to potassium channels (K^+-channel). SMS1 deficiency decreased KCNQ1 expression, which is a potential dependent K^+-channel in marginal cells, resulting in hearing impairment caused by the reduction of EP. In addition, oxidative stress was also involved in this down-regulation of KCNQ1 expression in SMS1-KO mice.

Recently, Kasahara et al. showed that SMS1-KO mice revealed the suppression of clot retraction in platelets [135]. Clot retraction is progressed by connection between fibrin fiber and actomyosin on the platelets membrane microdomain. Suppression of clot retraction in SMS1-KO seemed to be caused by the inhibition of the connection between the fibrin fiber and actomyosin due to the decrease of SM in microdomains. Because clot formation is an origin of thrombus, SMS1-dependen regulation of SM in platelets lipid microdomain may become a therapeutic target for thrombotic diseases.

Finally, Okazaki's group has described additional SMS1-KO mice phenotypes: 1) thrombocytopenia including a decrease of platelet numbers and a prolongation of bleeding time in SMS1-KO and 2) infertility in both male and female with shrinkage of testis and enlargement of uterus (unpublished data). Recently, SM in lipid microdomain was involved in the virus entry as well as ligand-receptor signaling as the infection of SMS1-KO mice by Japanese encephalitis virus was markedly reduced compared with wild type mice (unpublished data).

Sphingomyelin Synthase 2

SMS2-KO mice show no obvious abnormalities such as body weight loss and they have normal growth to adulthood with normal range of blood analysis. However, SMS2-KO revealed a decrease in SM contents and an increase in ceramides in various organs: macrophages (SM, 21% down; ceramide, 20% up), plasma (SM, 25% down; ceramide, 43% up), brain (SM, 30% down; ceramide, 39% up), and liver (SM, 25% down; ceramide 12% up) [143, 142]. In SMS2-KO mice, LPS-induced activation of NF-κB in macrophages and LPS-induced lung injury was mitigated [136, 137]. Moreover, the deterioration of atherosclerosis in the aorta of $Ldlr^{-/-}$ and ApoE KO mice ($ApoE^{-/-}$) was rescued by the transplantation with SMS2-KO mice-derived bone marrow cells and by mating with SMS2-KO mice, respectively, suggesting the implication of SMS2 in inflammation [138, 139]. Indeed, SMS2 deficiency suppressed the production of serum IL-6 and TNF-α after the LPS treatment.

Zhang et al. reported that SMS2-KO mice exhibited P-glycoprotein (Pgp, also named multi-drug resistant 1 (MDR1)) down-regulation in brain, which is a well-known drug transporter [140]. In addition, Mitsutake et al. showed that SMS2-KO inhibited diabetic phenomena such as a significant weight gain, the increase of blood glucose and triglyceride (TG), and liver steatosis in high fat diet (HFD)-mediated diabetic model [141]. Li et al. confirmed that SMS2-KO recovers reduction of insulin sensitivity and obesity induced by HFD [142]. The SMS2 deficiency decreased functions of fatty acid transporter CD36/FAT and caveolin 1, resulting in the suppression of fatty acid in liver. These results suggested that SM generated by SMS2 is related to the regulation of transporters or receptors on the microdomain of plasma membrane. On the other hand, Li et al. also reported the inhibition of HDF-induced liver steatosis in SMS2-KO mice [142, 188] showing a downregulation in peroxisome proliferator-activated receptor γ (PPAR-γ) expression, which results in the suppression of the lipid accumulation followed by steatosis. However, the authors link the effect on PPAR-γ expression to the increase in ceramide rather than to decrease in SM associated to the SMS2-deficiency. PPAR-γ is reported to transcriptionally regulate CD36 expression and it is involved in adipocyte differentiation and fat metabolism.

SMS2-KO mice showed an increase of infiltration of lymphoma EL-4 to liver in mouse infiltration models (unpublished data from Okazaki's group). Here, SMS2-mediated SM deficiency was involved in the attachment of EL-4 cells to liver endothelial cells. CPE activity in SMS2 KO mice will be mentioned in next section "SMSr."

Sphingomyelin Synthase Related

As it was mentioned, SMSr has no SMS- but CPE activity adding phosphoethanolamine to ceramide. Intracellular CPE cell levels are very low. Thus, it is thought that CPE production is not SMSr main role, but to control ceramide homeostasis in endoplasmic reticulum by its CPE activity, regulating so cell survival [55, 125]. However, CPE and SMSr biogenic function was unknown because SMSr-KO mice were not established until recently. In 2015, Ding et al. reported SMSr-KO and SMSr/SMS2 double knockout mice [56]. At the same time, Bickert et al. showed SMSr KO and knock-in mice mutated in active site of CPE synthesis [143].

CPE is ubiquitously distributed, and SMSr mRNA is also detected in various organs. However, the CPE amount in mice is only 1/300 of SM [143]. Interestingly, deficiency only in SMSr had no effect on macrophages, plasma, and liver CPE and ceramide cell levels. However, SMSr/SMS2-KO decreased CPE contents in plasma and liver significantly but not completely. In addition, SMS1-KO also decreased plasma CPE levels [56], suggesting that all SMSs members have activity of CPE synthesis, but there was no contribution of CPE synthesis to ceramide homeostasis in mice.

On the other hand, Bickert et al. found that SMSr is strongly expressed in brain and that CPE synthesis occurs, but no changes in the ceramide amount was detected. In addition, mutations in CPE active site of SMSr and SMS2 have no effects on mice CPE contents, development and fertility. From these results, the biogenic roles of CPE and CPE synthesis remain unclear.

E.5. Sphingomyelinase KO-Models

A-SMase

A-SMase is localized in lysosomes and it is characterized by acidic pH optima for enzymatic activity. A-SMase is ubiquitously expressed in mice. Two isoforms of the enzyme exist, one localized in lysosomes/endosomes (L-A-SMase) and other secreted form (S-A-SMase). At first, A-SMase is expressed as Zn^{2+}-independent L-A-SMase form, and then modified and released as Zn^{2+}-dependent S-A-SMase [144, 145]. As mentioned above, A-SMase is the responsible gene for NPD type A/B. A-SMase-KO exhibited SM accumulation and ER-stress injury in various organs such as liver, spleen, lung, and central nerves [146-149]. In this disease, excessive amounts of SM are accumulated in lysosomes because of A-SMase deficiency, which disrupts cellular organelles. A-SMase is activated through various stresses such as TNF-α, irradiation, and anti-cancer drugs with induction of apoptosis through ceramide generation. Santana et al. reported that treatment by X-irradiation to A-SMase-KO mice increased ceramide content in lungs and inhibited apoptosis in thymic lymphocytes [150].

Finally, Osawa et al. showed that in hepatic metastatic models of SL4 colon cancer, the lack of A-SMase enhanced tumor growth and inversely inhibited macrophages tumor infiltration [151]. This progressive effect of A-SMase deficiency on tumor growth was suppressed by induction of A-SMase in liver, suggesting that A-SMase activity has anti-tumor effects. These results showed that the well-established A-SMase regulation of the ceramide/SM cycle in cultured cells also works in mice.

N-SMase

N-SMase is Mg^{2+}-dependent and works in neutral conditions, pH 7.5. As already mentioned, there are four isoforms such as N-SMase1, N-SMase2, N-SMase3, and MA-N-SMase. In 1998 N-SMase1 was cloned from homology search of bacterial SMase [61]. Sawai et al. reported that N-SMase1 has SMase activity in vitro, but overexpression of N-SMase1 in HEK293 or MOLT4 cells could not show SM metabolic change [152]. Rather, N-SMase1 had phospholipase C activity on lysophosphatidylcholine or lyso-platelet activation factor (lyso-PAF) both in vitro and in the cells. Then, Zumbansen et al. established N-SMase1-KO mice and showed no obvious phenotype [153]. Particularly, N-SMase1 KO revealed no changes in SM and ceramide levels despite that N-SMase1 deficiency decreased SMase activity in vitro in different tissues (kidney, heart, and lung lysate). Moreover, there were no effects on LPC or lyso-PAF in N-SMase1 KO mice, disagreeing with the fact that N-SMase1 could act as the lysophospholipase C in vivo.

N-SMase2 is the most investigated isoform within the N-SMase family. Many reports show that N-SMase2 ceramide generation is involved in inflammation, apoptosis, and exosome production [154]. In 2005, Stoffel et al. reported that N-SMase2-KO mice show remaining N-SMase activity in liver and brain [155]. Then, they established N-SMase1 and N-SMase2 double KO mice (N-SMase1/N-SMase2-KO) [155]. N-SMase1/N-SMase2 exhibited complete suppression of N-SMase activity in brain, liver, and kidney. However, the changes in the main sphingolipids, including SM and ceramide, were not observed in these tissues. Finally, N-SMase2-KO mice show dwarfism-like phenotype including a clear growth delay and skeletal abnormality. Because N-SMase2 is involved in the vesicle transport from Golgi to extracellular site, it is thought that N-SMase2 deficiency suppresses the release of the growth factor from a hypothalamus neurosecretion neuron, resulting in dwarfism-like phenotype. At the same time, Aubin et al. found the deletion of 33 amino acids at C-terminal of N-SMase2 in *fragilitas ossium* (fro/fro) mice, which exhibits severe osteodystrophy [156]. In addition, osteoblast-specific transduction of N-SMase2 in fro/fro mice recovered dwarfism and osteodystrophy, suggesting that sphingolipids regulated by N-SMase2 are involved in bone mineralization [157, 158]. However, fro/fro mice, which are only 33 amino acids deletion of N-SMase2, showed more severe osteodystrophy than KO mice despite the fact that N-SMase2 KO mice had completely lost the N-SMase2 protein expression. It is possible that N-SMase2 has more unknown function to inhibit dwarfism or osteodystrophy independently of SMase activity. N-SMase3 was found from homology search of mammalian N-SMase1 and N-SMase2. Its expression is ubiquitous in human, and specifically strong in skeletal muscle and heart [63, 64]. N-SMase3 over-expression in MCF7 cells showed the enhancement of SMase activity by TNF-α treatment. Moreover, the increase of N-SMase activity occurred by TNF- α stimulation, and it was blocked by knockdown of N-SMase3, suggesting the implication of N-SMase3 in TNF- α signaling pathway [159]. However, N-SMase3-KO mice have not established yet, and its function as the regulator of sphingolipids *in vivo* remains unknown.

In 2010, Wu et al. found MA-SMase from homology search of zebrafish mitochondria-localized SMase [64]. Mouse MA-N-SMase has mitochondria localization signal and it is detected in mitochondria. SMase activity of MA-N-SMase is detected *in vitro*, but human orthologue of MA-N-SMase has no mitochondria signal. Unfortunately, MA-SMase-KO mice have not been established yet.

Alk-SMase

Alk-SMase seems to be involved in intestinal digestion of dietary SM, hydrolyzing it at optimal pH 9.0, into ceramide [160, 161]. In 2011, Zhang et al. reported that Alk-SMase-KO mice showed defective SMase activity in intestinal mucosa [162]. Hydrolysis of radiolabelled dietary SM to ceramide was suppressed up to 98% in Alk-SMase-KO mice. Inversely, in these mice, the residual radiolabelled-SM levels in intestinal mucosa showed 6-fold increase as compared with that in wild type. Interestingly, wild type mice showed conversion of palmitic acid to triglyceride in intestinal mucosa, but Alk-SMase-KO mice lost this phenomenon. Alk-SMase KO mice also showed 2.4-fold increase of labelled SM and 0.7-fold decrease of labelled ceramide in feces. Moreover, Alk-SMase was related with mucosal growth or with the maintenance of intestinal function such as alkaline phosphatase activity. These results suggested that Alk-SMase has a role in the digestion of the dietary SM in the intestinal mucosa. Alk-SMase-KO mice suppressed intestinal absorption of dietary cholesterol, suggesting that the Alk-SMase regulates intestinal absorption of SM and cholesterol in small intestine epithelium [163]. In addition, Alk-SMase was reported to be involved in cancer progression because the activity of Alk-SMase was reduced in human colon cancer or familial colitis [164, 165]. Indeed, in azoxymethane/dextran sulfate sodium (AOM/DSS) mice model for the study of colon carcinogenesis, Alk-SMase KO mice enhanced 4.5-fold in the number and 1.8-fold in the size of tumors compared with wild type mice [166]. Ceramide contents of small intestine and colon of Alk-SMase KO micewere markedly reduced, and accordingly an increase of Sph-1-P, which is one of pro-proliferative lipids, was observed. These results suggested that Alk-SMase has a tumor suppressive effect.

E.6. Role of SM in Brain Cells: Lessons from the Acid Sphingomyelinase Knock out Mouse

The remarkable enrichment of SM in neuronal membranes, together with the observation that most sphingolipidosis cause severe neurological alterations, argues in favor of a relevant role of this lipid in neuronal physiology. Taking into account that SM is a key element vesicle trafficking and that the latter is essential for the neuronal communication the severe phenotypes associated with alteration in SM metabolism are not surprising. However, research on this role has been challenging. The generation of mouse models in which SM metabolism has been genetically altered is helping to understand the molecular mechanisms in which brain SM is involved. Among these models is the mouse lacking the gene encoding for the A-SMase-KO (Horinouchi et al. 1995). The A-SMase-KO mouse recapitulates the symptoms of Niemann Pick disease type A caused by loss of function mutations in the ASM gene [167]. These include severe neurological alterations leading to motor and cognitive impairment, neurodegeneration and early death [149, 168, 169].

Several reasons make of the A-SMase-KO mouse a suitable system to study the role of SM in different neuronal types and subcellular compartments. First, alteration of SM levels appears to be quite specific. Thus, SM increase was confirmed in cultured primary hippocampal and cerebellar granule neurons from A-SMase-KO mice compared to wild type. However, levels of other lipids such as cholesterol, ceramide, triglycerides or gangliosides are not altered [170, 171]. Second, ASM deficiency results in the accumulation of SM not only in the lysosomes, where this enzyme mainly resides, but also at the plasma and synaptic membranes [170, 172]. Morphologically, the absence of a-SMase leads to mice having less synaptic vesicles and smaller presynaptic buttons [172]. On the other hand, since ASM is ubiquitous the analysis of A-SMase-KO mouse provides insight on the role of brain SM in non-neuronal cells. Next we describe recent studies in the brains of A-SMase-KO mice highlighting the important contribution of SM to neuronal survival and function as well as to glial cell physiology.

Vulnerability of Purkinje Cells in the Cerebellum of A-SMase-KO Mice: Role of SM in Neuronal Survival

A hallmark in A-SMase-KO mice brains is the rapid and complete degeneration of Purkinje cells of the cerebellum that accounts for motor impairment [169]. Interestingly, a subset of Purkinje cells, those zebrin II-negative, is particularly and very early compromised [173]. One of the possibilities proposed to explain this evidence relies on the role of plasma membrane SM in neuronal signaling. SM is a major component of raft lipid domains that are considered signaling platforms [174]. As such, SM affects the prosurvival signaling via p75 [175], which is especially prominent in Zebrin II- negative Purkinje cells [176].

Therefore, SM alterations could underlie p75 impaired signaling in A-SMase-KO zebrin negative Purkinje cells affecting their survival capacity. Supporting this, and other potential SM-mediated signaling alterations, is the high SM content of detergent insoluble membranes in A-SMase-KO brains, which are considered a biochemical correlate of raft lipid domains [170, 171]. SM increase may not only affect the distribution, activity and interactions of membrane proteins but also the biophysical properties of the membrane itself. In fact, the higher detergent-to-protein ratio needed to isolate A-SMase-KO with respect to wild type rafts likely reflects a reduced fluidity in restricted membrane areas [171].

Altered Axonal Molecular Polarization in A-SMase-KO Hippocampal Neurons: Role of SM in Endocytosis

Molecular polarization of axons and dendrites in neurons is essential for their function. Up regulation of SM synthesis during hippocampal neuron development is required to establish axonal polarization of certain molecules enriched in raft domains [177]. In addition, proper SM levels are necessary to keep their axonal distribution in the mature neuron [178]. The aberrantly high levels of SM in the plasma membrane of A-SMase-KO hippocampal neurons is responsible for the appearance in dendrites of GPI-anchored proteins such as the

Prion protein, which show a predominantly axonal distribution in control neurons [170]. The molecular mechanism underlying these alterations implies the influence of SM on endocytosis. GPI-anchored proteins are transported to both axons and dendrites once they are synthesized. Their final distribution restricted to the axons is acquired thanks to the rapid internalization from the dendritic surface. SM accumulation in the plasma membrane of A-SMase-KO neurons impairs GPI-anchored protein endocytosis by affecting the membrane attachment and activation of the small GTPase RhoA [170]. Consistently, reduction of SM levels by treatment with exogenous sphingomyelinase or expression of a RhoA constitutively active form restores endocytosis and axonal distribution of raft molecules in A-SMase-KO neurons [170]. These data unveiled the prominent role of SM in neuronal molecular trafficking.

Impaired Autophagy in A-SMase-KO Neurons: Role of SM in Lysosomal Membrane Permeabilization

Aberrantly high levels of SM in A-SMase-KO neurons occur both at the plasma membrane and the lysosomes. This contributes to the impairment of the autophagic flux. Indeed, autophagic vacuoles containing non-degraded material fill the cytosol of Purkinje cells in the cerebellum of A-SMase-KO mice [179]. This is not due to alterations in autophagy initiation or autophagosome-lysosome fusion but to the inefficient autophago-lysosomal clearance. High SM levels are responsible for the later since they induce lysosomal membrane permeabilization leading to the release of the lysosomal hydrolase Cathepsin B into the cytosol [179]. This reduces the capacity of autophagolysosomes to degrade their cargo and themselves. These results support a relevant role for SM in lysosomal membrane stability and in the modulation of autophagy in neurons.

Altered Presynaptic Plasticity in A-SMase-KO Neurons: Role of the SM Derivative Sphingosine in Synaptic Vesicle Docking

Degeneration of axonal synaptic terminals is among the earliest pathological events in A-SMase-KO neurons [169]. SM and its derivative sphingosine significantly accumulate in their synaptic membranes [172]. Electrophysiological recordings in hippocampal slices from A-SMase-KO mice indicated alterations in presynatic plasticity events such as paired pulse facilitation and post-tetanic potentiation, which were compatible with impaired neurotransmitter release. The reduced number of docked synaptic vesicles in A-SMase-KO presynaptic terminals contributes to these functional anomalies [172]. The underlying molecular mechanism does not directly involve SM but its derivative sphingosine. High levels of the later change the conformation of the SNARE syntaxin 1 altering its interaction with the cytosolic docking molecule Munc 18 [172]. Consistently, sphingosine reduces vesicle docking in cultured primary neurons and alters paired-pulse facilitation when added to hippocampal slices of wild type mice [172]. These evidences highlight the importance of modulating the levels of sphingosine and its precursor SM to preserve presynaptic plasticity.

Smaller Dendritic Spine Size in A-SMase-KO Neurons: Role of SM in Actin Polymerization

High levels of SM were also found in the postsynaptic membranes of A-SMase-KO neurons together with a smaller size of dendritic spines. These are protrusions of the postsynaptic membrane that receive most of the excitatory input in the central nervous system (CNS). Dynamic changes in the size and morphology of dendritic spines underlie memory and learning processes causing cognitive impairment when altered [180].

A recent study in A-SMase-KO mice identified SM as a key modulator of dendritic spine actin cytoskeleton, which is responsible for their size. SM accumulation in A-SMase-KO neurons decreases the levels of metabotropic glutamate receptors type I (mGluR1/5) at the postsynaptic membrane impairing membrane attachment and activity of the small GTPases RhoA and of its downstream effectors: the kinase ROCK and the actin binding protein Profilin IIa [181]. Impairment of this pathway results in reduced amount of polymerized actin that, in turn, accounts for the smaller size of A-SMase-KO dendritic spines. Pharmacological activation with dexamethasone of the neutral sphingomyelinase, an analogue enzyme to ASM that is present in synaptic membranes, prevents SM increase in A-SMase-KO neurons and enhances the amount of filamentous actin and the size of dendritic spines *in vitro* and *in vivo*. This strategy also has functional consequences improving cognitive and motor abilities of A-SMase-KO mice orally treated with dexamethasone [181]. These evidences confer to SM and its catabolic enzymes an important role in the regulation of actin polymerization in dendritic spines. They also indicate the influence of this lipid in neurotransmitter receptor physiology paving the way to study the underlying molecular mechanisms.

Altered Myelination and Immune Response in A-SMase-KO Mice Brains: Role of SM in Non-Neuronal Brain Cells

Oligodendrocytes are the myelin producing cells in the CNS. SM is a major component of the myelin sheet that insulates axons allowing the propagation of the nervous input. A time course study of the protein and mRNA expression of oligodendrocyte specific proteins involved in myelin architecture and function was carried out in A-SMase-KO mice. This study ruled out that ASM deficiency affects the proliferation and differentiation of oligodendrocyte precursors [182]. In contrast, significant alterations in myelin protein levels were observed in mature oligodendrocytes, which could explain the hypomyelination in these mice [182].

Although microglia have been considered the immune cells of the CNS, responsible for the release of cytokines, astrocytes have been recently shown to participate in inflammatory events. SM hydrolysis at the plasma membrane is essential for microparticle formation and IL-1b release from astrocytes [184]. The mechanism proposed involves changes in membrane fluidity. The SMase-induced reduction in SM content destabilizes the membrane thus facilitating blebbing and microparticle release [183]. Indeed, IL-1b release from astrocytes lacking ASM, which translocates to the plasma membrane to mediate microparticle shedding, is completely blocked [184]. This might explain the reduced levels of IL-1β found in the brain of A-SMase-KO mice [185] and the increased frequency of infections observed in Niemann Pick disease type A patients [186].

The above reported findings highlight the important contribution of A-SMase-KO mice to the understanding of SM roles in brain cell physiology and pathology. By influencing membrane fluidity, structure and compartmentalization or as precursor of derivatives like sphingosine, SM is involved in processes such as signaling, lysosomal membrane permeabilization, synaptic vesicle docking, actin polymerization, myelin integrity or cytokine release. Maintenance of proper levels of SM and its derivatives is necessary to ensure autophagy, synaptic plasticity, neuronal survival and immune response in the CNS. Playing such relevant roles the regulation of SM metabolic enzymes becomes an interesting target to fight against neurological diseases in which the levels of sphingolipids are altered.

CONCLUSION

Cells certainly invest a large amount of energy to maintain its lipid composition which is not only cell dependent, but within cell type, it varies according to it pathophysiological state. This lipid species diversity and their specific distribution opens two questions which, in fact, are common to all lipid classes: 1) what is the physiological meaning of these differences and 2) which are the mechanisms regulating, so strictly, the distribution of the thousands of existing molecular species. The traditional answer to the first question would link the lipid composition to the influence on the membrane biophysical properties. However, taking into account the existing complexity in regards of composition, this answer seems not to be sufficient or at least not specific enough.

To answer these questions we need to have a profound knowledge on synthetic pathways and all mechanisms regulating them, including substrate and product transport specificity, gene expression and protein activity regulation. Once more this is a common black box for most of lipids and for sure future studies will reveal the existence of new enzymes (or isoenzymes of known enzymes) because this seems to be the only way to obtain this highly specific diversity in composition. Examples of this are the six isoforms of ceramide synthases [35] or the fifteen of lysophosphatidyl acyl transferases [187] already described. These enzymes need to be specific in double sense: to the target (different sphingoid base or lysophospholipids) and to the different fatty acids.

Activities of SMS and SMase were already found in 1958 and 1967, respectively. A-SMase and N-SMase were cloned in 1989 and 1998, respectively. However, SMS genes have not been discovered for a long time. Only, after entering 21st century, in 2004, the genes coding for SMS (Sgms1, Sgms2, and Samd8) were cloned. On the other hand, in 2010, MA-N-SMase gene (Smpd5) was cloned at last. However, judging from the results of sphingolipids profile in each or combined KO mice the existing enzymes so far cannot explain all of *in vivo* regulation of sphingolipids such as SM and ceramide contents. For example, SM content is not completely reduced even in SMS1 and SMS2-KO condition (data not shown), suggesting that unknown SMS and SMase or alternative pathways to regulate SM and ceramide may exist. In addition, the functions of the different SM- and ceramide-molecular species remain to be investigated.

Sphingolipid metabolizing/catabolizing enzymes such as SMS and SMase and their regulation are involved in diverse functions including cell growth, death, differentiation, and autophagy in cell systems as well as animal and human systems. In the future, the application of animal disease model into SMS-KO and SMase-KO mice model may shed light on the real pathophysiological functions of SM and ceramide in vivo and the development of novel therapies targeting these enzymes and sphingolipids.

Finally, it is very important to keep in mind that the three aspects that SM seems clear to regulate: membrane properties, synthesis of bioactive molecules and vesicle formation, might be, and most probably are, highly connected.

ACKNOWLEDGMENTS

We are grateful to Dr. Jose Andrés Fernández (University of the Basque Country UPV/ EHU) for donating unpublished colon images and to Rebeca Reigada for helping during the manuscript edition. This study was partially supported by the Institute of Health Carlos III (Ministerio de Economía y Competitividad) and the EC (European Regional Development Fund, ERDF) (CP12/03338). GBC' contract was supported by the Miguel Servet program of the Institute of Health Carlos III. We thank the Ministerio Español de Ciencia e Innovación (SAF2011-24550 and CSD2010-00045), Fundación Niemann Pick España and Wylder Nation Foundation for the support to MDL. This work was also supported by: the Takeda Science Foundation (2012); a joint research project with Sharome Co., Ltd. (2012-2015); ONO Pharmaceutical Co., Ltd. (2012-2014); the JSPS KAKENHI Grant Number 25870853 (to MT); a grant from Strategic Research Foundation Grant-aided Project for Private Universities from Ministry of Education, Culture, Sport, Science, and Technology (MEXT) (No. S1201004; H2012-16); SENSHIN Medical Research Foundation; Mizutani Foundation for Glycoscience (2013-14); the Collaborative Research from Kanazawa Medical University (C2012-4, C2013-1). We are grateful to Dr. Satoshi Asano (Hiroshima University), Dr. Abo Bakr Abdel Shakor (King Khalid University), and Dr. Mayumi Hashimoto-Nishimura (Tottori University) to complete this work.

REFERENCES

[1] G. van Meer, D. R. Voelker, G. W. Feigenson, Membrane lipids: where they are and how they behave, *Nat. Rev. Mol. Cell. Biol.*, 9 (2008) 112-124.

[2] C. R. Gault, L. M. Obeid, Y. A. Hannun, An overview of sphingolipid metabolism: from synthesis to breakdown, *Adv. Exp. Med. Biol.*, 688 (2010) 1-23.

[3] J. Jin, Q. Hou, T. D. Mullen, Y. H. Zeidan, J. Bielawski, J. M. Kraveka, A. Bielawska, L. M. Obeid, Y. A. Hannun, Y. T. Hsu, Ceramide generated by sphingomyelin hydrolysis and the salvage pathway is involved in hypoxia/reoxygenation-induced Bax redistribution to mitochondria in NT-2 cells, *J. Biol. Chem.*, 283 (2008) 26509-26517.

[4] D. Milhas, C. J. Clarke, Y. A. Hannun, Sphingomyelin metabolism at the plasma membrane: implications for bioactive sphingolipids, *FEBS Lett.*, 584 (2010) 1887-1894.

[5] J. P. Slotte, Molecular properties of various structurally defined sphingomyelins -- correlation of structure with function, *Prog. Lipid Res.*, 52 (2013) 682.

[6] G. Barcelo-Coblijn, J. A. Fernandez, Mass spectrometry coupled to imaging techniques: the better the view the greater the challenge, *Frontiers in physiology*, 6 (2015) 3.

[7] F. X. Contreras, A. M. Ernst, P. Haberkant, P. Bjorkholm, E. Lindahl, B. Gonen, C. Tischer, A. Elofsson, G. von Heijne, C. Thiele, R. Pepperkok, F. Wieland, B. Brugger, Molecular recognition of a single sphingolipid species by a protein's transmembrane domain, *Nature*, 481 (2011) 525-529.

[8] J. Garate, R. Fernandez, S. Lage, J. Bestard-Escalas, D. H. Lopez, R. Reigada, S. Khorrami, D. Ginard, J. Reyes, I. Amengual, G. Barcelo-Coblijn, J. A. Fernandez, Imaging mass spectrometry increased resolution using 2-mercaptobenzothiazole and

2,5-diaminonaphtalene matrices: application to lipid distribution in human colon, *Analytical and bioanalytical chemistry*, 407 (2015) 4697-4708.

[9] K. Simons, E. Ikonen, Functional rafts in cell membranes, *Nature*, 387 (1997) 569-572.

[10] K. Simons, D. Toomre, Lipid rafts and signal transduction, *Nat. Rev. Mol. Cell. Biol.*, 1 (2000) 31-39.

[11] A. Filippov, G. Oradd, G. Lindblom, Sphingomyelin structure influences the lateral diffusion and raft formation in lipid bilayers, *Biophys. J.*, 90 (2006) 2086-2092.

[12] A. K. Gupta, H. Rudney, Plasma membrane sphingomyelin and the regulation of HMG-CoA reductase activity and cholesterol biosynthesis in cell cultures, *J. Lipid Res.*, 32 (1991) 125-136.

[13] J. P. Slotte, E. L. Bierman, Depletion of plasma-membrane sphingomyelin rapidly alters the distribution of cholesterol between plasma membranes and intracellular cholesterol pools in cultured fibroblasts, *Biochem. J.*, 250 (1988) 653-658.

[14] J. P. Slotte, Biological functions of sphingomyelins, *Prog. Lipid Res.*, 52 (2013) 424-437.

[15] M. Abe, A. Makino, F. Hullin-Matsuda, K. Kamijo, Y. Ohno-Iwashita, K. Hanada, H. Mizuno, A. Miyawaki, T. Kobayashi, A role for sphingomyelin-rich lipid domains in the accumulation of phosphatidylinositol-4,5-bisphosphate to the cleavage furrow during cytokinesis, *Mol. Cell. Biol.*, 32 (2012) 1396-1407.

[16] A. Othman, C. H. Saely, A. Muendlein, A. Vonbank, H. Drexel, A. von Eckardstein, T. Hornemann, Plasma 1-deoxysphingolipids are predictive biomarkers for type 2 diabetes mellitus, *BMJ Open Diabetes Res. Care*, 3 (2015) e000073.

[17] A. Othman, C. H. Saely, A. Muendlein, A. Vonbank, H. Drexel, A. von Eckardstein, T. Hornemann, Plasma C20-Sphingolipids predict cardiovascular events independently from conventional cardiovascular risk factors in patients undergoing coronary angiography, *Atherosclerosis*, 240 (2015) 216-221.

[18] H. Hama, Fatty acid 2-Hydroxylation in mammalian sphingolipid biology, *Biochim. Biophys. Acta*, 1801 (2010) 405-414.

[19] B. Zhou, J. F. Xiao, L. Tuli, H. W. Ressom, LC-MS-based metabolomics, *Molecular bioSystems*, 8 (2012) 470-481.

[20] E. Yasugi, T. Kasama, Y. Seyama, Composition of long chain bases in ceramide of the guinea pig Harderian gland, *Journal of biochemistry*, 110 (1991) 202-206.

[21] B. S. Robinson, D. W. Johnson, A. Poulos, Novel molecular species of sphingomyelin containing 2-hydroxylated polyenoic very-long-chain fatty acids in mammalian testes and spermatozoa, *J. Biol. Chem.*, 267 (1992) 1746-1751.

[22] J. K. Kramer, B. A. Blackwell, M. E. Dugan, F. D. Sauer, Identification of a new sphingolipid 3-O-acyl-D-erythro-sphingomyelin in newborn pig and infant plasma, *Biochim. Biophys. Acta*, 1303 (1996) 47-55.

[23] A. Bosio, E. Binczek, W. Stoffel, Functional breakdown of the lipid bilayer of the myelin membrane in central and peripheral nervous system by disrupted galactocerebroside synthesis, *Proc. Natl. Acad. Sci. US*, 93 (1996) 13280-13285.

[24] M. Valsecchi, L. Mauri, R. Casellato, S. Prioni, N. Loberto, A. Prinetti, V. Chigorno, S. Sonnino, Ceramide and sphingomyelin species of fibroblasts and neurons in culture, *J. Lipid Res.*, 48 (2007) 417-424.

[25] G. M. Oresti, J. M. Luquez, N. E. Furland, M. I. Aveldano, Uneven distribution of ceramides, sphingomyelins and glycerophospholipids between heads and tails of rat spermatozoa, *Lipids*, 46 (2011) 1081-1090.

[26] R. Sandhoff, Very long chain sphingolipids: tissue expression, function and synthesis, *FEBS Lett.*, 584 (2010) 1907-1913.

[27] N. O. Favale, B. J. Santacreu, L. G. Pescio, M. G. Marquez, N. B. Sterin-Speziale, Sphingomyelin metabolism is involved in the differentiation of MDCK cells induced by environmental hypertonicity, *J. Lipid Res.*, 56 (2015) 786-800.

[28] K. Kitatani, J. Idkowiak-Baldys, Y. A. Hannun, The sphingolipid salvage pathway in ceramide metabolism and signaling, *Cell Signal*, 20 (2008) 1010-1018.

[29] T. Okazaki, R. M. Bell, Y. A. Hannun, Sphingomyelin turnover induced by vitamin D3 in HL-60 cells. Role in cell differentiation, *J. Biol. Chem.*, 264 (1989) 19076-19080.

[30] Y. A. Hannun, L. M. Obeid, Principles of bioactive lipid signalling: lessons from sphingolipids, *Nat. Rev. Mol. Cell. Biol.*, 9 (2008) 139-150.

[31] M. M. Young, M. Kester, H. G. Wang, Sphingolipids: regulators of crosstalk between apoptosis and autophagy, *J. Lipid Res.*, 54 (2013) 5-19.

[32] M. E. Miller, S. Adhikary, A. A. Kolokoltsov, R. A. Davey, Ebolavirus requires acid sphingomyelinase activity and plasma membrane sphingomyelin for infection, *J. Virol.*, 86 (2012) 7473-7483.

[33] A. Othman, R. Bianchi, I. Alecu, Y. Wei, C. Porretta-Serapiglia, R. Lombardi, A. Chiorazzi, C. Meregalli, N. Oggioni, G. Cavaletti, G. Lauria, A. von Eckardstein, T. Hornemann, Lowering plasma 1-deoxysphingolipids improves neuropathy in diabetic rats, *Diabetes*, 64 (2015) 1035-1045.

[34] J. W. Park, W. J. Park, A. H. Futerman, Ceramide synthases as potential targets for therapeutic intervention in human diseases, *Biochim. Biophys. Acta*, 1841 (2014) 671-681.

[35] Y. Pewzner-Jung, S. Ben-Dor, A. H. Futerman, When do Lasses (longevity assurance genes) become CerS (ceramide synthases)?: Insights into the regulation of ceramide synthesis, *J. Biol. Chem.*, 281 (2006) 25001-25005.

[36] Y. A. Hannun, L. M. Obeid, Many ceramides, *J. Biol. Chem.*, 286 (2011) 27855-27862.

[37] B. J. Kroesen, B. Pettus, C. Luberto, M. Busman, H. Sietsma, L. de Leij, Y. A. Hannun, Induction of apoptosis through B-cell receptor cross-linking occurs via de novo generated C16-ceramide and involves mitochondria, *J. Biol. Chem.*, 276 (2001) 13606-13614.

[38] S. Koybasi, C. E. Senkal, K. Sundararaj, S. Spassieva, J. Bielawski, W. Osta, T. A. Day, J. C. Jiang, S. M. Jazwinski, Y. A. Hannun, L. M. Obeid, B. Ogretmen, Defects in cell growth regulation by C18:0-ceramide and longevity assurance gene 1 in human head and neck squamous cell carcinomas, *J. Biol. Chem.*, 279 (2004) 44311-44319.

[39] J. Mesicek, H. Lee, T. Feldman, X. Jiang, A. Skobeleva, E. V. Berdyshev, A. Haimovitz-Friedman, Z. Fuks, R. Kolesnick, Ceramide synthases 2, 5, and 6 confer distinct roles in radiation-induced apoptosis in HeLa cells, *Cell Signal*, 22 (2010) 1300-1307.

[40] T. D. Mullen, R. W. Jenkins, C. J. Clarke, J. Bielawski, Y. A. Hannun, L. M. Obeid, Ceramide synthase-dependent ceramide generation and programmed cell death: involvement of salvage pathway in regulating postmitochondrial events, *J. Biol. Chem.*, 286 (2011) 15929-15942.

[41] W. Zheng, J. Kollmeyer, H. Symolon, A. Momin, E. Munter, E. Wang, S. Kelly, J. C. Allegood, Y. Liu, Q. Peng, H. Ramaraju, M. C. Sullards, M. Cabot, A. H. Merrill, Jr., Ceramides and other bioactive sphingolipid backbones in health and disease: lipidomic analysis, metabolism and roles in membrane structure, dynamics, signaling and autophagy, *Biochim. Biophys. Acta*, 1758 (2006) 1864-1884.

[42] K. Venkataraman, A. H. Futerman, Ceramide as a second messenger: sticky solutions to sticky problems, *Trends Cell Biol.*, 10 (2000) 408-412.

[43] M. Fukasawa, M. Nishijima, K. Hanada, Genetic evidence for ATP-dependent endoplasmic reticulum-to-Golgi apparatus trafficking of ceramide for sphingomyelin synthesis in Chinese hamster ovary cells, *J. Cell Biol.*, 144 (1999) 673-685.

[44] A. H. Futerman, Y. A. Hannun, The complex life of simple sphingolipids, *EMBO Rep.*, 5 (2004) 777-782.

[45] K. Hanada, K. Kumagai, S. Yasuda, Y. Miura, M. Kawano, M. Fukasawa, M. Nishijima, Molecular machinery for non-vesicular trafficking of ceramide, *Nature*, 426 (2003) 803-809.

[46] A. H. Futerman, H. Riezman, The ins and outs of sphingolipid synthesis, *Trends Cell Biol.*, 15 (2005) 312-318.

[47] G. D'Angelo, E. Polishchuk, G. Di Tullio, M. Santoro, A. Di Campli, A. Godi, G. West, J. Bielawski, C. C. Chuang, A. C. van der Spoel, F. M. Platt, Y. A. Hannun, R. Polishchuk, P. Mattjus, M. A. De Matteis, Glycosphingolipid synthesis requires FAPP2 transfer of glucosylceramide, *Nature*, 449 (2007) 62-67.

[48] K. Huitema, J. van den Dikkenberg, J. F. Brouwers, J. C. Holthuis, Identification of a family of animal sphingomyelin synthases, *Embo J.*, 23 (2004) 33-44.

[49] A. van Helvoort, W. van't Hof, T. Ritsema, A. Sandra, G. van Meer, Conversion of diacylglycerol to phosphatidylcholine on the basolateral surface of epithelial (Madin-Darby canine kidney) cells. Evidence for the reverse action of a sphingomyelin synthase, *J. Biol. Chem.*, 269 (1994) 1763-1769.

[50] F. G. Tafesse, P. Ternes, J. C. Holthuis, The multigenic sphingomyelin synthase family, *J. Biol. Chem.*, 281 (2006) 29421-29425.

[51] S. Yamaoka, M. Miyaji, T. Kitano, H. Umehara, T. Okazaki, Expression cloning of a human cDNA restoring sphingomyelin synthesis and cell growth in sphingomyelin synthase-defective lymphoid cells, *J. Biol. Chem.*, 279 (2004) 18688-18693.

[52] P. Ternes, J. F. Brouwers, J. van den Dikkenberg, J. C. Holthuis, Sphingomyelin synthase SMS2 displays dual activity as ceramide phosphoethanolamine synthase, *J. Lipid Res.*, 50 (2009) 2270-2277.

[53] S. Asano, K. Kitatani, M. Taniguchi, M. Hashimoto, K. Zama, S. Mitsutake, Y. Igarashi, H. Takeya, J. Kigawa, A. Hayashi, H. Umehara, T. Okazaki, Regulation of Cell Migration by Sphingomyelin Synthases: Sphingomyelin in Lipid Rafts Decreases Responsiveness to Signaling by the Cxcl12/Cxcr4 Pathway, *Mol. Cell. Biol.*, (2012).

[54] A. B. Shakor, M. Taniguchi, K. Kitatani, M. Hashimoto, S. Asano, A. Hayashi, K. Nomura, J. Bielawski, A. Bielawska, K. Watanabe, T. Kobayashi, Y. Igarashi, H. Umehara, H. Takeya, T. Okazaki, Sphingomyelin synthase 1-generated sphingomyelin plays an important role in transferrin trafficking and cell proliferation, *J. Biol. Chem.*, 286 (2011) 36053-36062.

[55] A. M. Vacaru, F. G. Tafesse, P. Ternes, V. Kondylis, M. Hermansson, J. F. Brouwers, P. Somerharju, C. Rabouille, J. C. Holthuis, Sphingomyelin synthase-related protein SMSr controls ceramide homeostasis in the ER, *J. Cell Biol.*, 185 (2009) 1013-1027.
[56] T. Ding, I. Kabir, Y. Li, C. Lou, A. Yazdanyar, J. Xu, J. Dong, H. Zhou, T. Park, M. Boutjdir, Z. Li, X. C. Jiang, All members in the sphingomyelin synthase gene family have ceramide phosphoethanolamine synthase activity, *J. Lipid Res.*, 56 (2015) 537-545.
[57] N. Bartke, Y. A. Hannun, Bioactive sphingolipids: metabolism and function, *J. Lipid Res.*, 50 Suppl. (2009) S91-96.
[58] P. B. Schneider, E. P. Kennedy, Sphingomyelinase in normal human spleens and in spleens from subjects with Niemann-Pick disease, *J. Lipid Res.*, 8 (1967) 202-209.
[59] F. M. Goni, Non-permanent proteins in membranes: when proteins come as visitors (Review), *Mol. Membr. Biol.*, 19 (2002) 237-245.
[60] L. E. Quintern, E. H. Schuchman, O. Levran, M. Suchi, K. Ferlinz, H. Reinke, K. Sandhoff, R. J. Desnick, Isolation of cDNA clones encoding human acid sphingomyelinase: occurrence of alternatively processed transcripts, *Embo J.*, 8 (1989) 2469-2473.
[61] S. Tomiuk, K. Hofmann, M. Nix, M. Zumbansen, W. Stoffel, Cloned mammalian neutral sphingomyelinase: functions in sphingolipid signaling?, *Proc. Natl. Acad. Sci. US*, 95 (1998) 3638-3643.
[62] K. Hofmann, S. Tomiuk, G. Wolff, W. Stoffel, Cloning and characterization of the mammalian brain-specific, Mg2+-dependent neutral sphingomyelinase, *Proc. Natl. Acad. Sci. US*, 97 (2000) 5895-5900.
[63] O. Krut, K. Wiegmann, H. Kashkar, B. Yazdanpanah, M. Kronke, Novel tumor necrosis factor-responsive mammalian neutral sphingomyelinase-3 is a C-tail-anchored protein, *J. Biol. Chem.*, 281 (2006) 13784-13793.
[64] B. X. Wu, V. Rajagopalan, P. L. Roddy, C. J. Clarke, Y. A. Hannun, Identification and characterization of murine mitochondria-associated neutral sphingomyelinase (MA-nSMase), the mammalian sphingomyelin phosphodiesterase 5, *J. Biol. Chem.*, 285 (2010) 17993-18002.
[65] A. Nilsson, The presence of spingomyelin- and ceramide-cleaving enzymes in the small intestinal tract, *Biochim. Biophys. Acta*, 176 (1969) 339-347.
[66] J. Wu, Y. Cheng, C. Palmberg, T. Bergman, A. Nilsson, R. D. Duan, Cloning of alkaline sphingomyelinase from rat intestinal mucosa and adjusting of the hypothetical protein XP_221184 in GenBank, *Biochim. Biophys. Acta*, 1687 (2005) 94-102.
[67] S. Furuya, J. Mitoma, A. Makino, Y. Hirabayashi, Ceramide and its interconvertible metabolite sphingosine function as indispensable lipid factors involved in survival and dendritic differentiation of cerebellar Purkinje cells, *J. Neurochem.*, 71 (1998) 366-377.
[68] E. M. Jung, R. D. Griner, R. Mann-Blakeney, W. B. Bollag, A potential role for ceramide in the regulation of mouse epidermal keratinocyte proliferation and differentiation, *The Journal of investigative dermatology*, 110 (1998) 318-323.
[69] K. Iwai, T. Kondo, M. Watanabe, T. Yabu, T. Kitano, Y. Taguchi, H. Umehara, A. Takahashi, T. Uchiyama, T. Okazaki, Ceramide increases oxidative damage due to inhibition of catalase by caspase-3-dependent proteolysis in HL-60 cell apoptosis, *J. Biol. Chem.*, 278 (2003) 9813-9822.

[70] M. Kawase, M. Watanabe, T. Kondo, T. Yabu, Y. Taguchi, H. Umehara, T. Uchiyama, K. Mizuno, T. Okazaki, Increase of ceramide in adriamycin-induced HL-60 cell apoptosis: detection by a novel anti-ceramide antibody, *Biochim. Biophys. Acta*, 1584 (2002) 104-114.

[71] M. Taniguchi, K. Kitatani, T. Kondo, M. Hashimoto-Nishimura, S. Asano, A. Hayashi, S. Mitsutake, Y. Igarashi, H. Umehara, H. Takeya, J. Kigawa, T. Okazaki, Regulation of autophagy and its associated cell death by "sphingolipid rheostat": reciprocal role of ceramide and sphingosine 1-phosphate in the mammalian target of rapamycin pathway, *J. Biol. Chem.*, 287 (2012) 39898-39910.

[72] Y. Uchida, A. D. Nardo, V. Collins, P. M. Elias, W. M. Holleran, De novo ceramide synthesis participates in the ultraviolet B irradiation-induced apoptosis in undifferentiated cultured human keratinocytes, *The Journal of investigative dermatology*, 120 (2003) 662-669.

[73] M. Watanabe, T. Kitano, T. Kondo, T. Yabu, Y. Taguchi, M. Tashima, H. Umehara, N. Domae, T. Uchiyama, T. Okazaki, Increase of nuclear ceramide through caspase-3-dependent regulation of the "sphingomyelin cycle" in Fas-induced apoptosis, *Cancer Res.*, 64 (2004) 1000-1007.

[74] O. Cuvillier, Sphingosine in apoptosis signaling, *Biochim. Biophys. Acta*, 1585 (2002) 153-162.

[75] B. Oskouian, J. D. Saba, Cancer treatment strategies targeting sphingolipid metabolism, *Adv. Exp. Med. Biol.*, 688 (2010) 185-205.

[76] O. Cuvillier, G. Pirianov, B. Kleuser, P. G. Vanek, O. A. Coso, S. Gutkind, S. Spiegel, Suppression of ceramide-mediated programmed cell death by sphingosine-1-phosphate, *Nature*, 381 (1996) 800-803.

[77] S. Spiegel, S. Milstien, Sphingosine-1-phosphate: an enigmatic signalling lipid, *Nat. Rev. Mol. Cell. Biol.*, 4 (2003) 397-407.

[78] A. Gomez-Munoz, Ceramide 1-phosphate/ceramide, a switch between life and death, *Biochim. Biophys. Acta*, 1758 (2006) 2049-2056.

[79] A. Gomez-Munoz, J. Y. Kong, B. Salh, U. P. Steinbrecher, Ceramide-1-phosphate blocks apoptosis through inhibition of acid sphingomyelinase in macrophages, *J. Lipid Res.*, 45 (2004) 99-105.

[80] C. R. Vieira, J. M. Munoz-Olaya, J. Sot, S. Jimenez-Baranda, N. Izquierdo-Useros, J. L. Abad, B. Apellaniz, R. Delgado, J. Martinez-Picado, A. Alonso, J. Casas, J. L. Nieva, G. Fabrias, S. Manes, F. M. Goni, Dihydrosphingomyelin impairs HIV-1 infection by rigidifying liquid-ordered membrane domains, *Chemistry and biology*, 17 (2010) 766-775.

[81] R. F. Irvine, Nuclear lipid signalling, *Nat. Rev. Mol. Cell. Biol.*, 4 (2003) 349-360.

[82] G. Cascianelli, M. Villani, M. Tosti, F. Marini, E. Bartoccini, M. V. Magni, E. Albi, Lipid microdomains in cell nucleus, *Mol. Biol. Cell*, 19 (2008) 5289-5295.

[83] P. P. Ruvolo, Intracellular signal transduction pathways activated by ceramide and its metabolites, *Pharmacol. Res.*, 47 (2003) 383-392.

[84] D. K. Breslow, S. R. Collins, B. Bodenmiller, R. Aebersold, K. Simons, A. Shevchenko, C. S. Ejsing, J. S. Weissman, Orm family proteins mediate sphingolipid homeostasis, *Nature*, 463 (2010) 1048-1053.

[85] A. Ballabio, V. Gieselmann, Lysosomal disorders: from storage to cellular damage, *Biochim. Biophys. Acta*, 1793 (2009) 684-696.

[86] T. Kolter, K. Sandhoff, Sphingolipid metabolism diseases, *Biochim. Biophys. Acta*, 1758 (2006) 2057-2079.

[87] T. Kolter, K. Sandhoff, Lysosomal degradation of membrane lipids, *FEBS Lett.*, 584 (2010) 1700-1712.

[88] E. B. Vitner, F. M. Platt, A. H. Futerman, Common and uncommon pathogenic cascades in lysosomal storage diseases, *J. Biol. Chem.*, 285 (2010) 20423-20427.

[89] K. A. Dressler, C. C. Kan, R. N. Kolesnick, Sphingomyelin synthesis is involved in adherence during macrophage differentiation of HL-60 cells, *J. Biol. Chem.*, 266 (1991) 11522-11527.

[90] A. Qureshi, M. Subathra, A. Grey, K. Schey, M. Del Poeta, C. Luberto, Role of sphingomyelin synthase in controlling the antimicrobial activity of neutrophils against Cryptococcus neoformans, *PLoS One*, 5 (2010) e15587.

[91] A. Olivera, N. E. Buckley, S. Spiegel, Sphingomyelinase and cell-permeable ceramide analogs stimulate cellular proliferation in quiescent Swiss 3T3 fibroblasts, *J. Biol. Chem.*, 267 (1992) 26121-26127.

[92] Y. Nagata, H. Kobayashi, M. Umeda, N. Ohta, S. Kawashima, P. S. Zammit, R. Matsuda, Sphingomyelin levels in the plasma membrane correlate with the activation state of muscle satellite cells, *J. Histochem. Cytochem.*, 54 (2006) 375-384.

[93] G. Barcelo-Coblijn, M. L. Martin, R. F. de Almeida, M. A. Noguera-Salva, A. Marcilla-Etxenike, F. Guardiola-Serrano, A. Luth, B. Kleuser, J. E. Halver, P. V. Escriba, Sphingomyelin and sphingomyelin synthase (SMS) in the malignant transformation of glioma cells and in 2-hydroxyoleic acid therapy, *Proc. Natl. Acad. Sci. US*, 108 (2011) 19569-19574.

[94] T. E. Merchant, P. W. de Graaf, B. D. Minsky, H. Obertop, T. Glonek, Esophageal cancer phospholipid characterization by 31P NMR, *NMR Biomed.*, 6 (1993) 187-193.

[95] E. Marien, M. Meister, T. Muley, S. Fieuws, S. Bordel, R. Derua, J. Spraggins, R. Van de Plas, J. Dehairs, J. Wouters, M. Bagadi, H. Dienemann, M. Thomas, P. A. Schnabel, R. M. Caprioli, E. Waelkens, J. V. Swinnen, Non-small cell lung cancer is characterized by dramatic changes in phospholipid profiles, International journal of cancer. *Journal international du cancer*, (2015).

[96] T. E. Merchant, P. Meneses, L. W. Gierke, W. Den Otter, T. Glonek, 31P magnetic resonance phospholipid profiles of neoplastic human breast tissues, *Br. J. Cancer*, 63 (1991) 693-698.

[97] R. G. Cutler, J. Kelly, K. Storie, W. A. Pedersen, A. Tammara, K. Hatanpaa, J. C. Troncoso, M. P. Mattson, Involvement of oxidative stress-induced abnormalities in ceramide and cholesterol metabolism in brain aging and Alzheimer's disease, *Proc. Natl. Acad. Sci. US*, 101 (2004) 2070-2075.

[98] F. Samad, K. D. Hester, G. Yang, Y. A. Hannun, J. Bielawski, Altered adipose and plasma sphingolipid metabolism in obesity: a potential mechanism for cardiovascular and metabolic risk, *Diabetes*, 55 (2006) 2579-2587.

[99] I. Petrache, V. Natarajan, L. Zhen, T. R. Medler, A. T. Richter, C. Cho, W. C. Hubbard, E. V. Berdyshev, R. M. Tuder, Ceramide upregulation causes pulmonary cell apoptosis and emphysema-like disease in mice, *Nature medicine*, 11 (2005) 491-498.

[100] M. L. Doria, C. Z. Cotrim, C. Simoes, B. Macedo, P. Domingues, M. R. Domingues, L. A. Helguero, Lipidomic analysis of phospholipids from human mammary epithelial and breast cancer cell lines, *J. Cell. Physiol.*, 228 (2013) 457-468.

[101] A. Llorente, T. Skotland, T. Sylvanne, D. Kauhanen, T. Rog, A. Orlowski, I. Vattulainen, K. Ekroos, K. Sandvig, Molecular lipidomics of exosomes released by PC-3 prostate cancer cells, *Biochim. Biophys. Acta*, 1831 (2013) 1302-1309.

[102] J. C. Klock, J. K. Pieprzyk, Cholesterol, phospholipids, and fatty acids of normal immature neutrophils: comparison with acute myeloblastic leukemia cells and normal neutrophils, *J. Lipid Res.*, 20 (1979) 908-911.

[103] U. S. Eggert, T. J. Mitchison, C. M. Field, Animal cytokinesis: from parts list to mechanisms, *Annual review of biochemistry*, 75 (2006) 543-566.

[104] R. Rappaport, *Cytokinesis in Animal Cells Cambridge*, UK: Cambridge University Press, 1996.

[105] R. Albertson, B. Riggs, W. Sullivan, Membrane traffic: a driving force in cytokinesis, *Trends Cell Biol.*, 15 (2005) 92-101.

[106] L. I. Strickland, D. R. Burgess, Pathways for membrane trafficking during cytokinesis, *Trends Cell Biol.*, 14 (2004) 115-118.

[107] J. M. Mullins, J. J. Biesele, Terminal phase of cytokinesis in D-98s cells, *J. Cell Biol.*, 73 (1977) 672-684.

[108] K. Emoto, M. Umeda, An essential role for a membrane lipid in cytokinesis. Regulation of contractile ring disassembly by redistribution of phosphatidylethanolamine, *J. Cell Biol.*, 149 (2000) 1215-1224.

[109] D. Saul, L. Fabian, A. Forer, J. A. Brill, Continuous phosphatidylinositol metabolism is required for cleavage of crane fly spermatocytes, *J. Cell Sci.*, 117 (2004) 3887-3896.

[110] S. Yoshida, S. Bartolini, D. Pellman, Mechanisms for concentrating Rho1 during cytokinesis, *Genes and development*, 23 (2009) 810-823.

[111] A. Makino, M. Abe, M. Murate, T. Inaba, N. Yilmaz, F. Hullin-Matsuda, T. Kishimoto, N. L. Schieber, T. Taguchi, H. Arai, G. Anderluh, R. G. Parton, T. Kobayashi, Visualization of the heterogeneous membrane distribution of sphingomyelin associated with cytokinesis, cell polarity, and sphingolipidosis, *FASEB J.*, 29 (2015) 477-493.

[112] G. E. Atilla-Gokcumen, U. S. Eggert, A comparative LC-MS based profiling approach to analyze lipid composition in tissue culture systems, *Methods Mol. Biol.*, 1232 (2015) 103-113.

[113] E. Muro, G. E. Atilla-Gokcumen, U. S. Eggert, Lipids in cell biology: how can we understand them better?, *Mol. Biol. Cell*, 25 (2014) 1819-1823.

[114] K. M. Yamada, S. Even-Ram, Integrin regulation of growth factor receptors, *Nature cell biology*, 4 (2002) E75-76.

[115] K. H. Martin, J. K. Slack, S. A. Boerner, C. C. Martin, J. T. Parsons, Integrin connections map: to infinity and beyond, *Science*, 296 (2002) 1652-1653.

[116] R. K. Assoian, M. A. Schwartz, Coordinate signaling by integrins and receptor tyrosine kinases in the regulation of G1 phase cell-cycle progression, *Current opinion in genetics and development*, 11 (2001) 48-53.

[117] W. Baron, L. Decker, H. Colognato, C. ffrench-Constant, Regulation of integrin growth factor interactions in oligodendrocytes by lipid raft microdomains, *Curr. Biol.*, 13 (2003) 151-155.

[118] M. A. del Pozo, N. B. Alderson, W. B. Kiosses, H. H. Chiang, R. G. Anderson, M. A. Schwartz, Integrins regulate Rac targeting by internalization of membrane domains, *Science*, 303 (2004) 839-842.

[119] A. Norambuena, M. A. Schwartz, Effects of integrin-mediated cell adhesion on plasma membrane lipid raft components and signaling, *Mol. Biol. Cell*, 22 (2011) 3456-3464.

[120] M. G. Marquez, N. B. Sterin-Speziale, Is DRM lipid composition relevant in cell-extracellular matrix adhesion structures?, *Cell adhesion and migration*, 2 (2008) 180-183.

[121] B. Baum, M. Georgiou, Dynamics of adherens junctions in epithelial establishment, maintenance, and remodeling, *J. Cell Biol.*, 192 (2011) 907-917.

[122] M. G. Marquez, F. L. Nieto, M. C. Fernandez-Tome, N. O. Favale, N. Sterin-Speziale, Membrane lipid composition plays a central role in the maintenance of epithelial cell adhesion to the extracellular matrix, *Lipids*, 43 (2008) 343-352.

[123] H. Grassme, E. Gulbins, B. Brenner, K. Ferlinz, K. Sandhoff, K. Harzer, F. Lang, T. F. Meyer, Acidic sphingomyelinase mediates entry of N. gonorrhoeae into nonphagocytic cells, *Cell*, 91 (1997) 605-615.

[124] H. Grassme, A. Riehle, B. Wilker, E. Gulbins, Rhinoviruses infect human epithelial cells via ceramide-enriched membrane platforms, *J. Biol. Chem.*, 280 (2005) 26256-26262.

[125] F. G. Tafesse, A. M. Vacaru, E. F. Bosma, M. Hermansson, A. Jain, A. Hilderink, P. Somerharju, J. C. Holthuis, Sphingomyelin synthase-related protein SMSr is a suppressor of ceramide-induced mitochondrial apoptosis, *J. Cell Sci.*, 127 (2014) 445-454.

[126] M. C. Fernandes, M. Cortez, A. R. Flannery, C. Tam, R. A. Mortara, N. W. Andrews, Trypanosoma cruzi subverts the sphingomyelinase-mediated plasma membrane repair pathway for cell invasion, *The Journal of experimental medicine*, 208 (2011) 909-921.

[127] N. Nogueira, Z. Cohn, Trypanosoma cruzi: mechanism of entry and intracellular fate in mammalian cells, *The Journal of experimental medicine*, 143 (1976) 1402-1420.

[128] A. Reddy, E. V. Caler, N. W. Andrews, Plasma membrane repair is mediated by Ca(2+)-regulated exocytosis of lysosomes, *Cell*, 106 (2001) 157-169.

[129] I. S. Okoye, S. M. Coomes, V. S. Pelly, S. Czieso, V. Papayannopoulos, T. Tolmachova, M. C. Seabra, M. S. Wilson, MicroRNA-containing T-regulatory-cell-derived exosomes suppress pathogenic T helper 1 cells, *Immunity*, 41 (2014) 89-103.

[130] M. Yano, K. Watanabe, T. Yamamoto, K. Ikeda, T. Senokuchi, M. Lu, T. Kadomatsu, H. Tsukano, M. Ikawa, M. Okabe, S. Yamaoka, T. Okazaki, H. Umehara, T. Gotoh, W. J. Song, K. Node, R. Taguchi, K. Yamagata, Y. Oike, Mitochondrial dysfunction and increased reactive oxygen species impair insulin secretion in sphingomyelin synthase 1-null mice, *J. Biol. Chem.*, 286 (2011) 3992-4002.

[131] M. Yano, T. Yamamoto, N. Nishimura, T. Gotoh, K. Watanabe, K. Ikeda, Y. Garan, R. Taguchi, K. Node, T. Okazaki, Y. Oike, Increased oxidative stress impairs adipose tissue function in sphingomyelin synthase 1 null mice, *PLoS One*, 8 (2013) e61380.

[132] Z. Li, Y. Fan, J. Liu, Y. Li, C. Huan, H. H. Bui, M. S. Kuo, T. S. Park, G. Cao, X. C. Jiang, Impact of sphingomyelin synthase 1 deficiency on sphingolipid metabolism and atherosclerosis in mice, *Arterioscler. Thromb. Vasc. Biol.*, 32 (2012) 1577-1584.

[133] L. Dong, K. Watanabe, M. Itoh, C. R. Huan, X. P. Tong, T. Nakamura, M. Miki, H. Iwao, A. Nakajima, T. Sakai, T. Kawanami, T. Sawaki, Y. Masaki, T. Fukushima, Y. Fujita, M. Tanaka, M. Yano, T. Okazaki, H. Umehara, CD4+ T-cell dysfunctions through the impaired lipid rafts ameliorate concanavalin A-induced hepatitis in sphingomyelin synthase 1-knockout mice, *Int. Immunol.*, 24 (2012) 327-337.

[134] M. H. Lu, M. Takemoto, K. Watanabe, H. Luo, M. Nishimura, M. Yano, H. Tomimoto, T. Okazaki, Y. Oike, W. J. Song, Deficiency of sphingomyelin synthase-1 but not sphingomyelin synthase-2 causes hearing impairments in mice, *J. Physiol.*, 590 (2012) 4029-4044.

[135] K. Kasahara, M. Kaneda, T. Miki, K. Iida, N. Sekino-Suzuki, I. Kawashima, H. Suzuki, M. Shimonaka, M. Arai, Y. Ohno-Iwashita, S. Kojima, M. Abe, T. Kobayashi, T. Okazaki, M. Souri, A. Ichinose, N. Yamamoto, Clot retraction is mediated by factor XIII-dependent fibrin-alphaIIbbeta3-myosin axis in platelet sphingomyelin-rich membrane rafts, *Blood*, 122 (2013) 3340-3348.

[136] S. Gowda, C. Yeang, S. Wadgaonkar, F. Anjum, N. Grinkina, M. Cutaia, X.C. Jiang, R. Wadgaonkar, Sphingomyelin synthase 2 (SMS2) deficiency attenuates LPS-induced lung injury, *Am. J. Physiol. Lung Cell. Mol. Physiol.*, 300 (2011) L430-440.

[137] T. K. Hailemariam, C. Huan, J. Liu, Z. Li, C. Roman, M. Kalbfeisch, H. H. Bui, D. A. Peake, M. S. Kuo, G. Cao, R. Wadgaonkar, X. C. Jiang, Sphingomyelin synthase 2 deficiency attenuates NFkappaB activation, *Arterioscler. Thromb. Vasc. Biol.*, 28 (2008) 1519-1526.

[138] Y. Fan, F. Shi, J. Liu, J. Dong, H. H. Bui, D. A. Peake, M. S. Kuo, G. Cao, X. C. Jiang, Selective reduction in the sphingomyelin content of atherogenic lipoproteins inhibits their retention in murine aortas and the subsequent development of atherosclerosis, *Arterioscler. Thromb. Vasc. Biol.*, 30 (2010) 2114-2120.

[139] J. Liu, C. Huan, M. Chakraborty, H. Zhang, D. Lu, M. S. Kuo, G. Cao, X. C. Jiang, Macrophage sphingomyelin synthase 2 deficiency decreases atherosclerosis in mice, *Circ. Res.*, 105 (2009) 295-303.

[140] Y. Zhang, J. Dong, X. Zhu, W. Wang, Q. Yang, The effect of sphingomyelin synthase 2 (SMS2) deficiency on the expression of drug transporters in mouse brain, *Biochem. Pharmacol.*, 82 (2011) 287-294.

[141] S. Mitsutake, K. Zama, H. Yokota, T. Yoshida, M. Tanaka, M. Mitsui, M. Ikawa, M. Okabe, Y. Tanaka, T. Yamashita, H. Takemoto, T. Okazaki, K. Watanabe, Y. Igarashi, Dynamic modification of sphingomyelin in lipid microdomains controls development of obesity, fatty liver, and type 2 diabetes, *J. Biol. Chem.*, 286 (2011) 28544-28555.

[142] Z. Li, H. Zhang, J. Liu, C. P. Liang, Y. Li, Y. Li, G. Teitelman, T. Beyer, H. H. Bui, D. A. Peake, Y. Zhang, P. E. Sanders, M. S. Kuo, T. S. Park, G. Cao, X. C. Jiang, Reducing plasma membrane sphingomyelin increases insulin sensitivity, *Mol. Cell. Biol.*, 31 (2011) 4205-4218.

[143] A. Bickert, C. Ginkel, M. Kol, K. vom Dorp, H. Jastrow, J. Degen, R. L. Jacobs, D. E. Vance, E. Winterhager, X. C. Jiang, P. Dormann, P. Somerharju, J. C. Holthuis, K. Willecke, Functional characterization of enzymes catalyzing ceramide phosphoethanolamine biosynthesis in mice, *J. Lipid Res.*, 56 (2015) 821-835.

[144] R. W. Jenkins, D. Canals, Y. A. Hannun, Roles and regulation of secretory and lysosomal acid sphingomyelinase, *Cell Signal.*, 21 (2009) 836-846.

[145] E. L. Smith, E. H. Schuchman, The unexpected role of acid sphingomyelinase in cell death and the pathophysiology of common diseases, *Faseb J.*, 22 (2008) 3419-3431.

[146] K. Horinouchi, S. Erlich, D. P. Perl, K. Ferlinz, C. L. Bisgaier, K. Sandhoff, R. J. Desnick, C. L. Stewart, E. H. Schuchman, Acid sphingomyelinase deficient mice: a model of types A and B Niemann-Pick disease, *Nat. Genet.*, 10 (1995) 288-293.

[147] G. Hua, R. Kolesnick, Using ASMase knockout mice to model human diseases, *Handb. Exp. Pharmacol.*, (2013) 29-54.

[148] B. Otterbach, W. Stoffel, Acid sphingomyelinase-deficient mice mimic the neurovisceral form of human lysosomal storage disease (Niemann-Pick disease), *Cell*, 81 (1995) 1053-1061.

[149] E. H. Schuchman, The pathogenesis and treatment of acid sphingomyelinase-deficient Niemann-Pick disease, *J. Inherit. Metab. Dis.*, 30 (2007) 654-663.

[150] P. Santana, L. A. Pena, A. Haimovitz-Friedman, S. Martin, D. Green, M. McLoughlin, C. Cordon-Cardo, E. H. Schuchman, Z. Fuks, R. Kolesnick, Acid sphingomyelinase-deficient human lymphoblasts and mice are defective in radiation-induced apoptosis, *Cell*, 86 (1996) 189-199.

[151] Y. Osawa, A. Suetsugu, R. Matsushima-Nishiwaki, I. Yasuda, T. Saibara, H. Moriwaki, M. Seishima, O. Kozawa, Liver acid sphingomyelinase inhibits growth of metastatic colon cancer, *J. Clin. Invest.*, 123 (2013) 834-843.

[152] H. Sawai, N. Domae, N. Nagan, Y. A. Hannun, Function of the cloned putative neutral sphingomyelinase as lyso-platelet activating factor-phospholipase C, *J. Biol. Chem.*, 274 (1999) 38131-38139.

[153] M. Zumbansen, W. Stoffel, Neutral sphingomyelinase 1 deficiency in the mouse causes no lipid storage disease, *Mol. Cell. Biol.*, 22 (2002) 3633-3638.

[154] A. A. Shamseddine, M. V. Airola, Y. A. Hannun, Roles and regulation of neutral sphingomyelinase-2 in cellular and pathological processes, *Adv. Biol. Regul.*, 57 (2015) 24-41.

[155] W. Stoffel, B. Jenke, B. Block, M. Zumbansen, J. Koebke, Neutral sphingomyelinase 2 (smpd3) in the control of postnatal growth and development, *Proc. Natl. Acad. Sci. US*, 102 (2005) 4554-4559.

[156] I. Aubin, C. P. Adams, S. Opsahl, D. Septier, C. E. Bishop, N. Auge, R. Salvayre, A. Negre-Salvayre, M. Goldberg, J. L. Guenet, C. Poirier, A deletion in the gene encoding sphingomyelin phosphodiesterase 3 (Smpd3) results in osteogenesis and dentinogenesis imperfecta in the mouse, *Nat. Genet.*, 37 (2005) 803-805.

[157] Z. Khavandgar, C. Poirier, C. J. Clarke, J. Li, N. Wang, M. D. McKee, Y. A. Hannun, M. Murshed, A cell-autonomous requirement for neutral sphingomyelinase 2 in bone mineralization, *J. Cell Biol.*, 194 (2011) 277-289.

[158] W. Stoffel, B. Jenke, B. Holz, E. Binczek, R. H. Gunter, J. Knifka, J. Koebke, A. Niehoff, Neutral sphingomyelinase (SMPD3) deficiency causes a novel form of chondrodysplasia and dwarfism that is rescued by Col2A1-driven smpd3 transgene expression, *Am. J. Pathol.*, 171 (2007) 153-161.

[159] J. S. Moylan, J. D. Smith, E. M. Wolf Horrell, J. B. McLean, G. M. Deevska, M. R. Bonnell, M. N. Nikolova-Karakashian, M. B. Reid, Neutral sphingomyelinase-3 mediates TNF-stimulated oxidant activity in skeletal muscle, *Redox Biol.*, 2 (2014) 910-920.

[160] R. D. Duan, A. Nilsson, Metabolism of sphingolipids in the gut and its relation to inflammation and cancer development, *Prog. Lipid Res.*, 48 (2009) 62-72.

[161] A. Nilsson, R. D. Duan, Absorption and lipoprotein transport of sphingomyelin, *J. Lipid Res.*, 47 (2006) 154-171.

[162] Y. Zhang, Y. Cheng, G. H. Hansen, L. L. Niels-Christiansen, F. Koentgen, L. Ohlsson, A. Nilsson, R. D. Duan, Crucial role of alkaline sphingomyelinase in sphingomyelin digestion: a study on enzyme knockout mice, *J. Lipid Res.*, 52 (2011) 771-781.

[163] P. Zhang, Y. Chen, Y. Cheng, E. Hertervig, L. Ohlsson, A. Nilsson, R. D. Duan, Alkaline sphingomyelinase (NPP7) promotes cholesterol absorption by affecting sphingomyelin levels in the gut: A study with NPP7 knockout mice, *Am. J. Physiol. Gastrointest. Liver Physiol.*, 306 (2014) G903-908.

[164] E. Hertervig, A. Nilsson, J. Bjork, R. Hultkrantz, R. D. Duan, Familial adenomatous polyposis is associated with a marked decrease in alkaline sphingomyelinase activity: a key factor to the unrestrained cell proliferation?, *Br. J. Cancer*, 81 (1999) 232-236.

[165] E. Hertervig, A. Nilsson, L. Nyberg, R. D. Duan, Alkaline sphingomyelinase activity is decreased in human colorectal carcinoma, *Cancer*, 79 (1997) 448-453.

[166] Y. Chen, P. Zhang, S. C. Xu, L. Yang, U. Voss, E. Ekblad, Y. Wu, Y. Min, E. Hertervig, A. Nilsson, R. D. Duan, Enhanced colonic tumorigenesis in alkaline sphingomyelinase (NPP7) knockout mice, *Mol. Cancer Ther.*, 14 (2014) 259-267.

[167] R. O. Brady, J. N. Kanfer, M. B. Mock, D. S. Fredrickson, The metabolism of sphingomyelin. II. Evidence of an enzymatic deficiency in Niemann-Pick diseae, *Proc. Natl. Acad. Sci. US*, 55 (1966) 366-369.

[168] M. D. Ledesma, A. Prinetti, S. Sonnino, E. H. Schuchman, Brain pathology in Niemann Pick disease type A: insights from the acid sphingomyelinase knockout mice, *J. Neurochem.*, 116 (2010) 779-788.

[169] S. L. Macauley, R. L. Sidman, E. H. Schuchman, T. Taksir, G. R. Stewart, Neuropathology of the acid sphingomyelinase knockout mouse model of Niemann-Pick A disease including structure-function studies associated with cerebellar Purkinje cell degeneration, *Exp. Neurol.*, 214 (2008) 181-192.

[170] C. Galvan, P. G. Camoletto, F. Cristofani, P. P. Van Veldhoven, M. D. Ledesma, Anomalous surface distribution of glycosyl phosphatidyl inositol-anchored proteins in neurons lacking acid sphingomyelinase, *Mol. Biol. Cell*, 19 (2008) 509-522.

[171] F. Scandroglio, J. K. Venkata, N. Loberto, S. Prioni, E. H. Schuchman, V. Chigorno, A. Prinetti, S. Sonnino, Lipid content of brain, brain membrane lipid domains, and neurons from acid sphingomyelinase deficient mice, *J. Neurochem.*, 107 (2008) 329-338.

[172] P. G. Camoletto, H. Vara, L. Morando, E. Connell, F. P. Marletto, M. Giustetto, M. Sassoe-Pognetto, P. P. Van Veldhoven, M. D. Ledesma, Synaptic vesicle docking: sphingosine regulates syntaxin1 interaction with Munc18, *PLoS One*, 4 (2009) e5310.

[173] J. Sarna, S. R. Miranda, E. H. Schuchman, R. Hawkes, Patterned cerebellar Purkinje cell death in a transgenic mouse model of Niemann Pick type A/B disease, *Eur. J. Neurosci.*, 13 (2001) 1873-1880.

[174] Y. Zhang, X. Li, K. A. Becker, E. Gulbins, Ceramide-enriched membrane domains--structure and function, *Biochim. Biophys. Acta*, 1788 (2009) 178-183.

[175] A. B. Brann, R. Scott, Y. Neuberger, D. Abulafia, S. Boldin, M. Fainzilber, A. H. Futerman, Ceramide signaling downstream of the p75 neurotrophin receptor mediates the effects of nerve growth factor on outgrowth of cultured hippocampal neurons, *J. Neurosci.*, 19 (1999) 8199-8206.

[176] I. Dusart, M. P. Morel, C. Sotelo, Parasagittal compartmentation of adult rat Purkinje cells expressing the low-affinity nerve growth factor receptor: changes of pattern expression after a traumatic lesion, *Neuroscience*, 63 (1994) 351-356.

[177] M. D. Ledesma, B. Brugger, C. Bunning, F. T. Wieland, C. G. Dotti, Maturation of the axonal plasma membrane requires upregulation of sphingomyelin synthesis and formation of protein-lipid complexes, *Embo J.*, 18 (1999) 1761-1771.

[178] C. Galvan, P. G. Camoletto, C. G. Dotti, A. Aguzzi, M. D. Ledesma, Proper axonal distribution of PrP(C) depends on cholesterol-sphingomyelin-enriched membrane domains and is developmentally regulated in hippocampal neurons, *Mol. Cell. Neurosci.*, 30 (2005) 304-315.

[179] E. Gabande-Rodriguez, P. Boya, V. Labrador, C. G. Dotti, M. D. Ledesma, High sphingomyelin levels induce lysosomal damage and autophagy dysfunction in Niemann Pick disease type A, *Cell Death Differ.*, 21 (2014) 864-875.

[180] R. Yuste, T. Bonhoeffer, Morphological changes in dendritic spines associated with long-term synaptic plasticity, *Annu. Rev. Neurosci.*, 24 (2001) 1071-1089.

[181] A. I. Arroyo, P. G. Camoletto, L. Morando, M. Sassoe-Pognetto, M. Giustetto, P. P. Van Veldhoven, E. H. Schuchman, M. D. Ledesma, Pharmacological reversion of sphingomyelin-induced dendritic spine anomalies in a Niemann Pick disease type A mouse model, *EMBO Mol. Med.*, 6 (2014) 398-413.

[182] B. Buccinna, M. Piccinini, A. Prinetti, F. Scandroglio, S. Prioni, M. Valsecchi, B. Votta, S. Grifoni, E. Lupino, C. Ramondetti, E. H. Schuchman, M. T. Giordana, S. Sonnino, M. T. Rinaudo, Alterations of myelin-specific proteins and sphingolipids characterize the brains of acid sphingomyelinase-deficient mice, an animal model of Niemann-Pick disease type A, *J. Neurochem.*, 109 (2009) 105-115.

[183] A. D. Tepper, P. Ruurs, T. Wiedmer, P. J. Sims, J. Borst, W. J. van Blitterswijk, Sphingomyelin hydrolysis to ceramide during the execution phase of apoptosis results from phospholipid scrambling and alters cell-surface morphology, *J. Cell Biol.*, 150 (2000) 155-164.

[184] F. Bianco, C. Perrotta, L. Novellino, M. Francolini, L. Riganti, E. Menna, L. Saglietti, E. H. Schuchman, R. Furlan, E. Clementi, M. Matteoli, C. Verderio, Acid sphingomyelinase activity triggers microparticle release from glial cells, *Embo J.*, 28 (2009) 1043-1054.

[185] C. G. Ng, D. E. Griffin, Acid sphingomyelinase deficiency increases susceptibility to fatal alphavirus encephalomyelitis, *J. Virol.*, 80 (2006) 10989-10999.

[186] O. A. Minai, E. J. Sullivan, J. K. Stoller, Pulmonary involvement in Niemann-Pick disease: case report and literature review, *Respiratory medicine*, 94 (2000) 1241-1251.

[187] H. Shindou, D. Hishikawa, T. Harayama, K. Yuki, T. Shimizu, Recent progress on acyl CoA: lysophospholipid acyltransferase research, *J. Lipid Res.*, 50 Suppl. (2009) S46-51.

[188] Y. Li, J. Dong, T. Ding, M. S. Kuo, G. Cao, X.C. Jiang, Z. Li Z, Sphingomyelin synthase 2 activity and liver steatosis: an effect of ceramide-mediated peroxisome proliferator-activated receptor γ2 suppression, *Arterioscler Thromb Vasc Biol.* 33 (2013) 1513-20.

Chapter 6

Unravelling the Role of Sphingosine-Based Lipids Using Sphingolipidomics

M. Rosário M Domingues, Ana Reis, Daniela Couto and Pedro Domingues

Mass Spectrometry Centre, Department of Chemistry and QOPNA, University of Aveiro, Campus Universitário de Santiago, Aveiro, Portugal

Abstract

Sphingolipids, a heterogeneous class of lipids, are key players as constituents of cell membranes, and as signalling molecules. Bioactive sphingolipids mediate many key cellular processes, including cell metabolism, growth, differentiation, stress and inflammatory responses, cell apoptosis and autophagy. Altered sphingolipids profiles have been frequently observed in various diseases, such as inflammatory and neurodegenerative diseases, cancer and sphingolipidoses. They are in particularly relevant in Gaucher and Fabry diseases.

The development of mass spectrometric methods and technologies has improved the comprehensive analysis of the total sphingolipid profile, and allows one to decipher their biological functions in health and disease. Based on different analytical approaches, sphingolipidomics has been used for the molecular characterization and quantification of sphingolipid classes and molecular species. This chapter describes the basis of modern mass spectrometry approaches to sphingolipidomics, and the application to the study of sphingolipids in living systems. Special emphasis is placed on disease states.

Keywords: sphingolipids, mass spectrometry, lipidomics, gangliosides

1. The Challenges Resulting from Structural Diversity of Sphingoid-Based Lipids

Sphingolipids (SLs) are a heterogeneous class of lipids. They are key players as signaling molecules and as constituents of cell membranes. In cell membranes, SLs are asymmetrically distributed across the lipid bilayer, and although they are predominantly located in the outer membrane leaflet, they can also be found in specific domains, known as lipid rafts [1]. These microdomains can influence the location and function of transporter and receptor proteins [2]. Bioactive sphingolipids mediate many key cellular processes, including cellular metabolism, growth, differentiation, stress and inflammatory responses, in addition to cell apoptosis and autophagy [3].

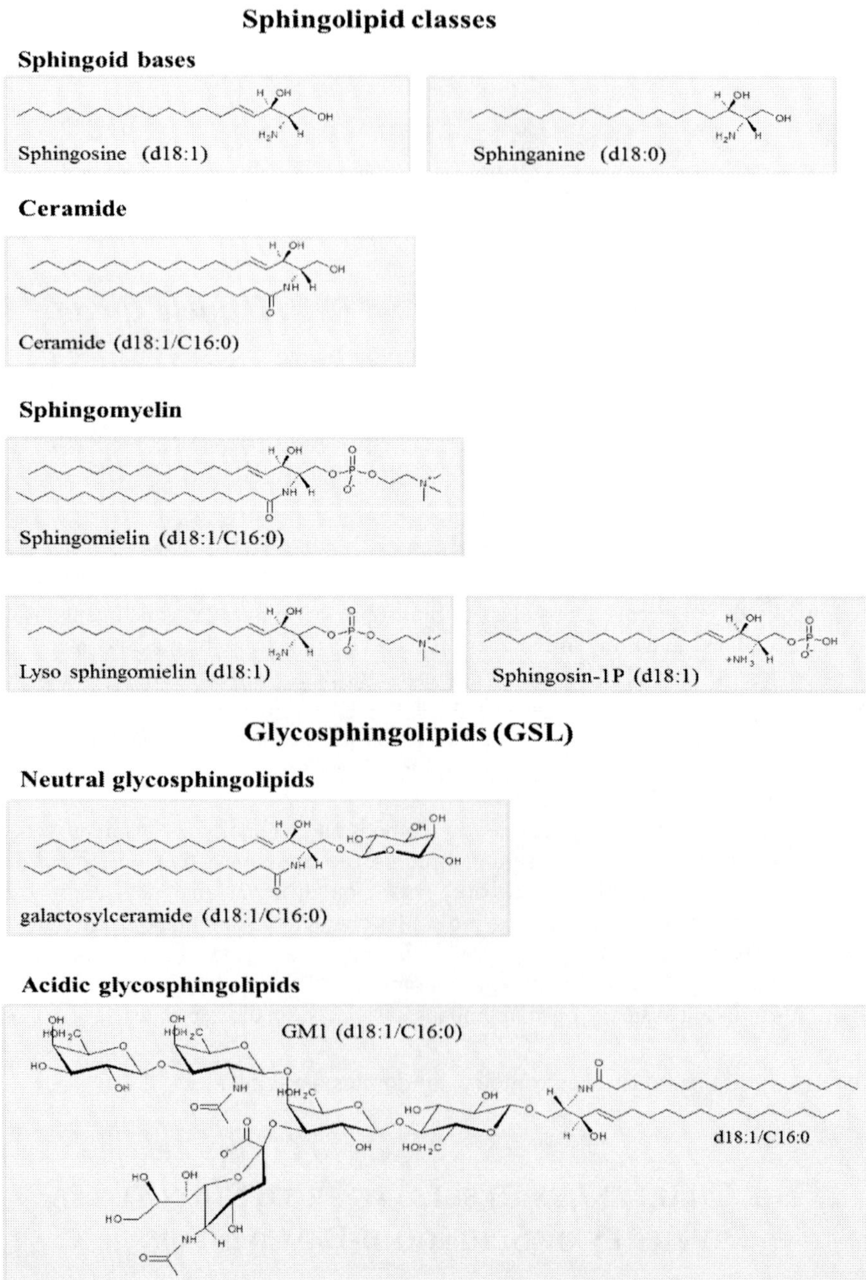

Figure 1. Classification and structure of the main classes of sphingolipids.

Sphingolipids have a common sphingoid base backbone, which may be linked to a fatty acyl chain, and different structural types of polar head group. The most common backbone is sphingosine (2-amino-4-octadecene-1,3-diol), an 18-carbon amino alcohol with an unsaturated hydrocarbon chain. Sphingolipids can be categorized as sphingomyelin (SM), ceramide (Cer), sphingosine-1-phosphate (S1P) and glycosphingolipids (GSLs), (Figure 1) [4, 5]. GSLs, which may be neutral and acidic, are a class which involves a wide variety of compounds. All the GSLs share a hydrophobic core, consisting of a skeleton of ceramide, and a carbohydrate moiety. Various, very different types of carbohydrate may be involved, which can range from glucose and galactose (glucosylceramide and galactosylceramide), to di, tri and tetrasaccharides, or, in the case of gangliosides, even more complex structures. In addition, the possibility of additional acetyl and sulfatide groups increases the structural complexity of GSLs. The GSLs bearing glucose oligosaccharides are called globosides, while those which have sialic acid are classified as gangliosides. The polar head groups of gangliosides may have one or more residues of the sialic acid N-acetyl neuraminic acid. This group is responsible for the negative charge of gangliosides at pH 7, which distinguishes them from globosides and cerebrosides, which only have neutral sugars (glucose, galactose or galactosamine) [6]. Gangliosides can be classified into different families, based on the number of sialic acid groups present in the polar head (Figure 1).

GSLs are particularly abundant in the central and peripheral nervous system. Galactosylceramide (GalCer) is the main component of myelin, while gangliosides are strategic components in lipid rafts where they, along with cholesterol and sphingomyelin, are involved in the regulation of cell surface signalling and neurotransmission [7, 8]. Both are important in maintaining homeostasis in the central nervous system, and of stabilization of the myelin and nerve regeneration [7, 8]. A large number of enzymes control this network of sphingolipids, regulating their metabolism and maintaining the appropriate ratio of the sphingolipid classes. The disturbance of the SL metabolism, which can be expressed in terms of changes in profile, has been associated with situations such as neurodegeneration [9], cancer [10], inflammation and immune diseases [11]. Of particular relevance to sphingolipidoses are Fabry and Gaucher diseases [12, 13].

Sphingolipidomics aims to identify and quantify all sphingolipid molecular species, and to unravel their correlation with genes and proteins to understand the function in living systems [14, 15]. Different analytical strategies have been used, to evaluate the homeostasis or deregulation of the metabolism of sphingolipids, including chromatographic methods, such as thin layer chromatography (TLC) [16–18], immunochemical methods [19], and hyphenated high pressure liquid chromatography-mass spectrometry-based approaches (HPLC-MS), using electrospray ionization (ESI-MS) [15, 20, 21] or matrix assisted laser desorption ionization (MALDI-MS) [15, 17, 20, 21]. Several lipidomic approaches can be used to characterize the sphingolipid profile, usually starting with the extraction of lipids from cells, tissues or biological fluids, followed by sample fractionation and sphingolipid detection. As in global lipidomics analysis, the methodological approach that is selected can have huge influence on the nature and interpretation of the data obtained. Different analytical approaches can, therefore, be used, depending on the objective of the study, such that the experimental design and methodology must be carefully planned.

The following section describes commonly used analytical methods that have been reported for quantitative and qualitative characterization of sphingolipids, with special emphasis on modern strategies using mass spectrometry. We will focus on liquid chromatography (LC) coupled to mass spectrometry and tandem mass spectrometry (MS/MS), using both untargeted and targeted approaches to identify the sphingolipidome profile of the biological samples.

2. Advanced Analytical Strategies Used in Sphingolipidomics

Sphingolipidomics is a multistage analytical approach based on three key steps: extraction of sphingolipids from cells, tissues or biological fluids, followed by the fractionation and detection of the different lipid classes and species, and, finally, data analysis, to attempt to identify and quantify the classes and molecular species of SLs (Figure 2) [15, 21]. The extraction procedure is a critical step in sphingolipidomics analysis. The lipidomics analysis, usually involves analysis of the organic phase obtained from an extraction protocol, based on methods such as those of Folch [22], Bligh and Dyer [23], or the MTBE method [24]. However, to obtain the total sphingolipidomic profiles, both organic and aqueous phases have to be analysed. The organic phase (lower phase) is rich in phospholipids, sphingomyelin, ceramides and cerebrosides (GalCer and GlcCer) and globoside, while gangliosides are extracted in the aqueous phase. Most published reports have only identified sphingolipids, such as SM, Cer, GlcCer or LacCer, extracted in the organic phase. The fractionation and detection of the different lipid classes is usually performed by LC-MS, using either reversed-phase chromatography [25] or, more recently, HILIC LC-MS [25–27]. In some studies, direct analysis of the samples using a shotgun lipidomics approach, has been reported [28]. However, less abundant molecular species of sphingolipids, in particular Cer and GalCer, may not be observed, due to the lower concentrations of sphingolipids relative to the phospholipids, which are also present in the organic phase. To overcome this drawback, mild alkaline hydrolysis can be used to hydrolyse the ester bonds of glycerolipids (triglycerides and phospholipids) [14, 21]. Following this, a new extraction using chloroform-methanol / water can be performed and the organic phase, enriched with sphingolipids, is then recovered.

Gangliosides are isolated from the aqueous phase after a Folch or Bligh and Dyer extraction. Following this, the aqueous phase is usually cleaned to remove non-lipid components by solid phase extraction (SPE), using a reversed-phase (RP) C18 cartridges, followed by mild alkaline hydrolysis and purification of gangliosides using anion exchange chromatography on DEAE-Sephadex [7, 8, 29]. After extraction, various approaches can be used to identify and quantify the sphingolipids. Thin layer chromatography remains one of the most commonly used methods for fractionating sphingolipids, in particular, GSLs. This technique allows the separation and identification of individual sphingolipid classes from complex mixtures. It does not allow the determination of the chemical structure of sphingolipids, but provides information about their possible nature, based on the chromatographic mobility and their reactivity with specific stain dyes [30].

Glycosphingolipids can be selectively stained in TLC plates with appropriate reagents. The spots of glycosphingolipids classes are usually identified as bluish bands using irreversible staining with resorcinol [31]. For quantification of individual ganglioside classes, the ganglioside being visualized is analysed by densitometric scanning using a 580 nm laser

[16, 32]. Spots of glycolipids can also be visualized, following treatment with a dilute solution of orcinol [17, 33]. Individual spots of any class of lipids can also be visualized in the TLC plate using either iodine vapour, or the reversible dye primuline. Primuline has enormous advantages, since lipid spots identified by it can be scratched from TLC plates, and further analysed by ESI-MS or MALDI-MS. The MALDI-MS approach has been used to detected sphingomyelin and ceramides in lipid extracts [34–36]. Direct TLC analysis by MALDI-MS can also be performed, allowing much faster data acquisition, but requires the use of specific inlets and TLC plates [37]. Direct interfacing of the TLC plate with desorption electrospray ionization DESI-MS has also been carried out [38]. More recently, TLC-immune blotting using specific antibodies to GSLs has become a more sensitive and specific alternative for their identification and quantification [39, 40]. Antibodies have been used to identify sub-population gangliosides, including in microarrays [41], as a tool for screening GSLs in biological fluids and tissues [41–43]. However, immune chemical approaches are still very expensive and there are some difficulties in obtaining highly pure and specific antibodies.

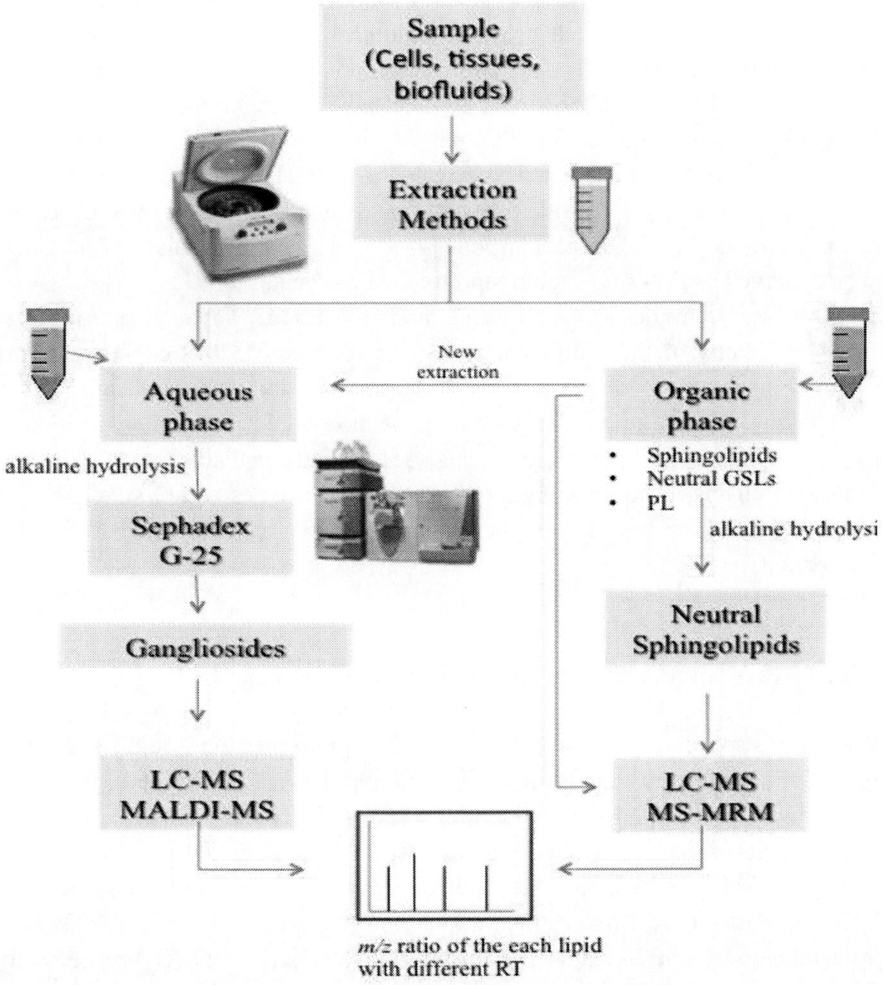

Figure 2. Sphingolipidomic workflows.

The new generation of user-friendly mass spectrometers with high scan rates and high resolution has allowed the extensive use of mass spectrometry for high-throughput analysis of biological samples and is now an exciting approach for searching sphingolipid signatures of disease in fluids, cells and tissues. By coupling liquid chromatography to mass spectrometry, it is possible to separate sphingolipid classes and/or species when using, respectively, normal phase or reversed phase columns. Details on the structural characteristics of the GSL can be obtained by analysis using tandem mass spectrometry (MS/MS). Untargeted or targeted lipidomic LC-MS based strategies can be performed to quantify lipid molecular species. All these approaches have been used to study the sphingolipidome of biological samples, as will be described in the next section.

3. MASS SPECTROMETRY-BASED LIPIDOMICS FOR SPHINGOLIPID ANALYSIS

Sphingolipids have been extensively studied using ESI-MS and MALDI-MS. In the MS spectra, neutral sphingolipids, such as SM, ceramides, GalCer and LacCer, are generally observed in positive mode as $[M + H]^+$, $[M+Na]^+$ and/or $[M + Li]^+$ molecular ions [44]. When using an LC-MS approach, SM can be detected in negative mode as $[M + CH_3COO]^-$ or $[M + HCOO]^-$, depending on the chromatographic additive that was used (ammonium acetate or formate). Ceramides can also be detected as $[M - H]^-$ molecular ions in the negative ion mode. Sulfatides and gangliosides carrying acid moieties, are usually observed in the negative ion mode, as $[M - H]^-$ molecular ions [45–47]. Higher molecular weight gangliosides can also be observed as $[M - nH]^{n-}$ multideprotonated molecular ions [21, 48].

Tandem mass spectrometry or multistage tandem MS (MS^n) analysis is usually required to confirm the identity of the individual molecular species. MS/MS experiments provide a well-developed approach to obtain structural information about the molecular species in a sample by taking advantage of the unique gas-phase molecular decomposition of each precursor ion. These mass spectrometric analyses provide valuable structural information about fatty acyl chains, head groups and chain sphingoid types of sphingolipids [49]. The typical fragmentation pathway of each class of sphingolipids will be described in detail in the following section.

a. Tandem Mass Spectrometry of Sphingoid-Base Lipids

Tandem mass spectra of sphingomyelin $[M + H]^+$ molecular ions yield a typical fragment ion at m/z 184, attributed to the phosphatidylcholine polar head group [50], which is similar to that seen with glycerophosphatidylcholine (PC) lipids [51]. However, in spite of the similarity of the gas-phase behavior of SM and PC, they can be easily distinguished in the MS spectra since the SM molecular ions have odd *m/z* values, while the PC ions have even ones. The MS/MS spectra of the SM negative adducts $[M + CH_3COO]^-$ or $[M + HCOO]^-$ show product ions due to loss of 74 Da ($-CH_3COOCH_3$) or 60 Da ($HCOOCH_3$) respectively, and a product ion at *m/z* 168, assigned to a demethylated polar head (Table 1).

Ceramides can be identified in tandem mass spectra, acquired in both positive and negative ion modes. The tandem mass spectra of protonated molecular ions of ceramides carrying a sphingosine backbone typically yield product ions at *m/z* 264 and *m/z* 282 (Table 1). However, ceramides containing dihydrosphingosine moieties and threo and erythro isoforms normally yield product ions at *m/z* 266 and 284 (Table 1) [52]. These are typical of the sphingosine with a chain of C_{18}, the most common naturally abundant one in sphingolipids. However, these m/z values may differ, depending on whether the sphinganine chain or the sphingosine acyl is present. Other common product ions observed in these MS/MS spectra, include those arising from loss of one or more water molecules, assigned as $[M + H - nH_2O]^+$ ions [52]. The MS/MS spectra of $[M\text{-}H]^-$ molecular ions of ceramides provide supplementary information about the molecular structure of the fatty acyl (N-acyl) and sphingoid chain. The loss of the fatty acyl moiety in ceramides may occur after cleavage of N - C = O, followed by elimination of the fatty acyl chain as ketene (-R = C = O), or of cleavage of the C_2-N and elimination of $RCONH_2$ [53, 54]. The tandem mass spectra of ceramides containing sphingosine show product ions arising from the neutral loss of 30 Da ($H_2C = O$), while ceramides containing sphinganine show product ions arising from the neutral loss of 48 Da (due to neutral loss of $H_2C = O + H_2O$) [53, 54].

The MS spectrum of phosphorylated sphingoid bases, S1P, is characterized by $[M + H]^+$ or $[M\text{-}H]^-$ molecular ions when analyzed in positive or negative ion mode, respectively [55]. The MS/MS spectrum of S1P in negative ion mode shows a characteristic abundant product ion at *m/z* 79.0, corresponding to $[PO_3]^-$ [55, 56], and a low abundant product ion at *m/z* 97, corresponding to $[H_2PO_4]^-$ [56]. Collision-induced dissociation of the protonated $[M + H]^+$ S1P molecular ion yields abundant product ions at *m/z* 264, arising from the combined neutral loss of H_3PO_4 and H_2O [56].

Neutral glycolipids, such as hexosylceramides (HexCer) and lactosylceramides, are usually analyzed in the positive mode, while the acidic gangliosides are preferably analyzed in negative mode. In MS/MS analysis, protonated ions from GlcCer, LacCer and more complex glycolipids, $[M + H]^+$, undergo dissociation through the following pathways:

a) the cleavage of glycosidic bonds with loss of the carbohydrate head group as a neutral species. In this case, the charge is retained in the ceramide moiety, and the product ion $[Cer + H]^+$ is observed [20].
b) at high collision energies, cleavage of both the carbohydrate head groups and the acyl chain of the fatty acid occurs. In this case, the charge is retained in the dehydrated sphingoid base, which is usually observed as product ions at *m/z* 264 or 266 [20]. The MS/MS spectra of cationized molecular ions, $[M + X]^+$, X = Na, Li, yield additional products ions of the sugar moieties, which allow confirmation of the sugar composition of glycolipids, as shown for C24:1GalCer and C24:1 LacCer in Figure 3 and 4 [45, 57]. GalCer (bearing a galactose residue - Figure 3) and LacCer (bearing a galactose-glucose disaccharide - Figure 4), are shown here as examples of neutral GSLs fragmentation patterns under MS/MS conditions. The MS/MS spectra of protonated GalCer and LacCer, show characteristic product ions arising from loss of sugar moieties: loss of Gal_{res} (-162 Da) [58] and of Gal (-180 Da) from the GalCer; loss of Lac_{res} (-324 Da), Lac (-342 Da), Glc_{res} (-162 Da) and of Glc (- 180 Da) from the LacCer. In both MS/MS spectra, diagnostic product ions, corresponding to sphingosine moiety, are observed at *m/z* 264 [45, 57]. The MS/MS spectra of

sodiated adducts, $[M + Na]^+$, of GalCer and LacCer show the same characteristic losses, in particular the neutral loss of sugar moiety, as observed for $[M + H]^+$ ions. Additional product ions, corresponding to the unmodified glycosidic moieties at *m/z* 185.1 $[Gal_{res} + Na]^+$, 203.1 $[Gal + Na]^+$, 347.2 $[Lac_{res} + Na]^+$ and 365.2 $[Lac + Na]^+$ are observed, as shown in Figures 3 and 4, [45, 57].

The MS/MS spectrum of deprotonated $[M\text{-}H]^-$ sulfatide molecular ions is dominated by the product ions at *m/z* 97 (HSO_4^-) and 80 (SO_3^-), confirming the presence of a sulfate group [45, 59]. The MS/MS spectra of sodiated $[M + Na]^+$ molecular ions show abundant product ions at *m/z* 143 $[NaHSO_4 + Na]^+$ and at *m/z* 264, due to $[\text{sphingosine-}2H_2O + H]^+$, which is a typical product ion of ceramides [59].

Gangliosides have very complex structures and, therefore, their fragmentation pathways are also complex. They can be analyzed either in positive or negative mode [48], although most of the published work reported the use in the negative mode.

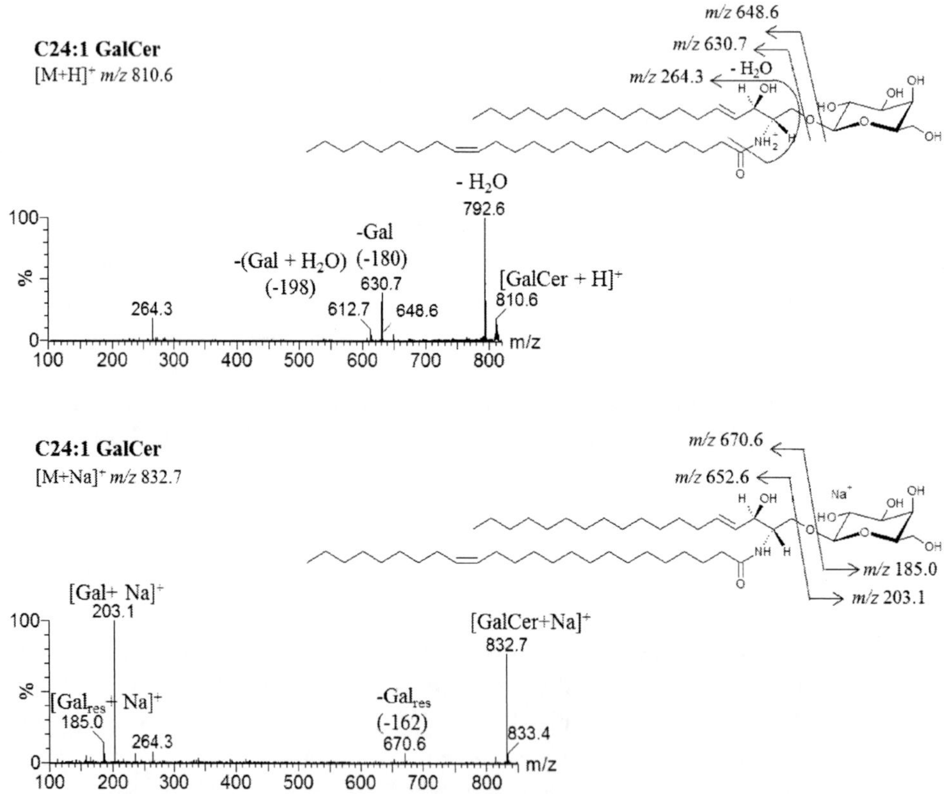

Figure 3. ESI-Q-TOF tandem mass spectra of C24:1GalCer $[M + H]^+$ at *m/z* 810.6 and $[M + Na]^+$ at *m/z* 832.7 (Reproduced and adapted from Santinha et al. 2014, Rapid Commun Mass Spectrom, 2014, with permission from Wiley. Copyright 2014, Wiley).

This mode takes advantage of the presence of the sialic acid moiety that increases the sensitivity of the analysis [48, 60]. Characteristic product ions in the MS/MS spectra of the gangliosides are those arising from neutral loss of sialic acid residues/N-acetyl neuraminic acid residues ($NeuAc_{res}$) (-291 Da) and the species at *m/z* 290 $[NeuAc_{res}\text{-}H]^-$ and at *m/z* 308

$[NeuAc–H]^-$ [48]. Product ions arising from the neutral loss of glycan with formation the product ions $[Cer-H]^-$ and $[Cer-30-H]^-$ are also typical of ganglioside MS/MS spectra [46]. The MS/MS spectra of the double charged and/or triple charged molecular ions can provide valuable structural information, by showing the glycan anion $[glycan-H]^-$ and $[glycan-H_2O-H]^-$; for example, Figure 5 shows the MS/MS spectrum of the $[M-2H]^{2-}$ molecular ion of GM1(d18:1/C18:0). The MS/MS spectra of gangliosides also shows product ions arising from loss of CO_2 (-44 Da), and those formed due to glycosidic cleavage in the glycan polar head [46, 48]. The identification of the typical product ions and neutral losses of each class sphingolipids is essential for defining target lipidomic strategies based on the precursor ion scan (PIS), neutral loss scan (NLS) or multiple reaction monitoring (MRM) approaches, as shown in Table 1.

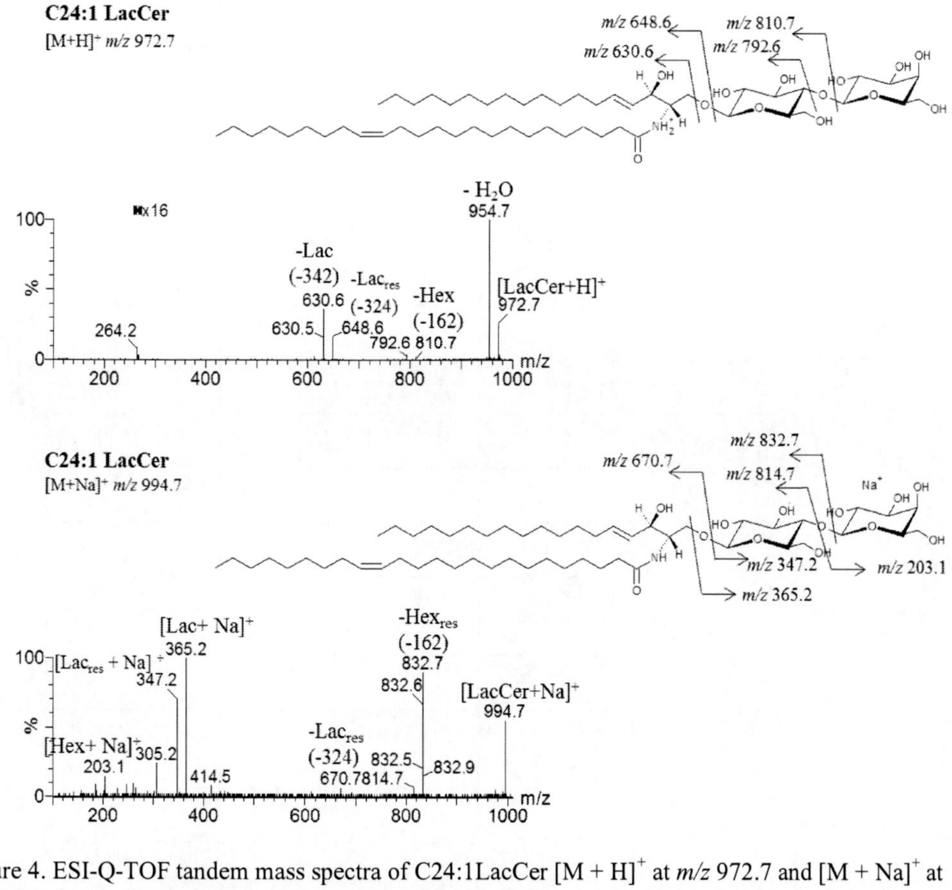

Figure 4. ESI-Q-TOF tandem mass spectra of C24:1LacCer $[M + H]^+$ at *m/z* 972.7 and $[M + Na]^+$ at m/z 994.7 (Reproduced and adapted from Santinha et al. 2014, Rapid Commun Mass Spectrom, 2014, with permission from Wiley. Copyright 2014, Wiley).

Table 1. Main neutral losses and product ions observed in the MS/MS spectra of the sphingolipid classes Cer, SM, S1P, GSL that can be used for target MS analysis by using neutral loss scanning (NLS) or precursor ion scanning (PIS) modes

		Neutral loss	Neutral loss scanning (NLS)	m/z	Product ion scanning (PIS)	Ref
Ceramides	$[M + H]^+$			264	[sphingosine-$2H_2O$ + $H]^+$	[47, 52, 111]
				266	[sphinganine-$2H_2O$ + $H]^+$	[52, 111]
	$[M-H]^-$	NL 32	CH_3OH			[53, 54]
		NL 30	$H_2C = O$			
Sphingosine-1P	$[M + H]^+$	NL 116	$H_3PO_4 + H_2O$	264	[sphingosine-$2H_2O$ + $H]^+$	[47, 56]
				266	[sphinganine-$2H_2O$ + $H]^+$	[53, 54, 111]
	$[M-H]^-$			97	$[H_2PO_4]^-$	[56]
				79	$[PO_3]^-$	[56]
Sulfatides	$[M-H]^-$			97	$[HSO_4]^-$	[21, 45,47, 59]
	$[M + Na]^+$			143	$[NaHSO_4 + Na]^+$	[59]
Sphingomyelin	$[M + H]^+$			184	[Phosphocholine + $H]^+$	[51, 111–113]
	$[M + HCOO]^-$	NL 60	$HCO_2 + CH_3$			[114]
	[M + CH_3COO]-	NL 74	CH_3COOCH_3			[114]
	$[M-H]^-$			168	[Phosphocholine-CH_3- $H]^-$	[51]
Glycosphingolipids: GalCer LacCer Globosides	$[M + H]^+$	NL 162	$C_6H_{10}O_5$			[115-117]
		NL 324 For Hex_nCer ($n \geq 2$)	$C_{12}H_{20}O_{10}$			[117]
		NL 180	$C_6H_{12}O_6$			[116, 117]
		NL 342 For Hex_nCer ($n \geq 2$)	$C_{12}H_{22}O_{11}$			[117]
	$[M + Na]^+$			185	$[Hex_{res} + Na]^+$	[117]
				347 For Hex_n Cer ($n \geq 2$)	$[Lac_{res} + Na]^+$	
				203	$[Hex + Na]^+$	
				365 For Hex_n Cer ($n \geq 2$)	$[Lac + Na]^+$	
Gangliosides	$[M-H]^-$	NL 291 [118]	$C_{11}H_{17}NO_8$	290	$[NeuAc_{res} - H]^-$	[21, 47]
	$[M-2H]^{2-}$	NL 291 [118]	$C_{11}H_{17}NO_8$	290	$[NeuAc_{res} - H]^-$	[21, 111]

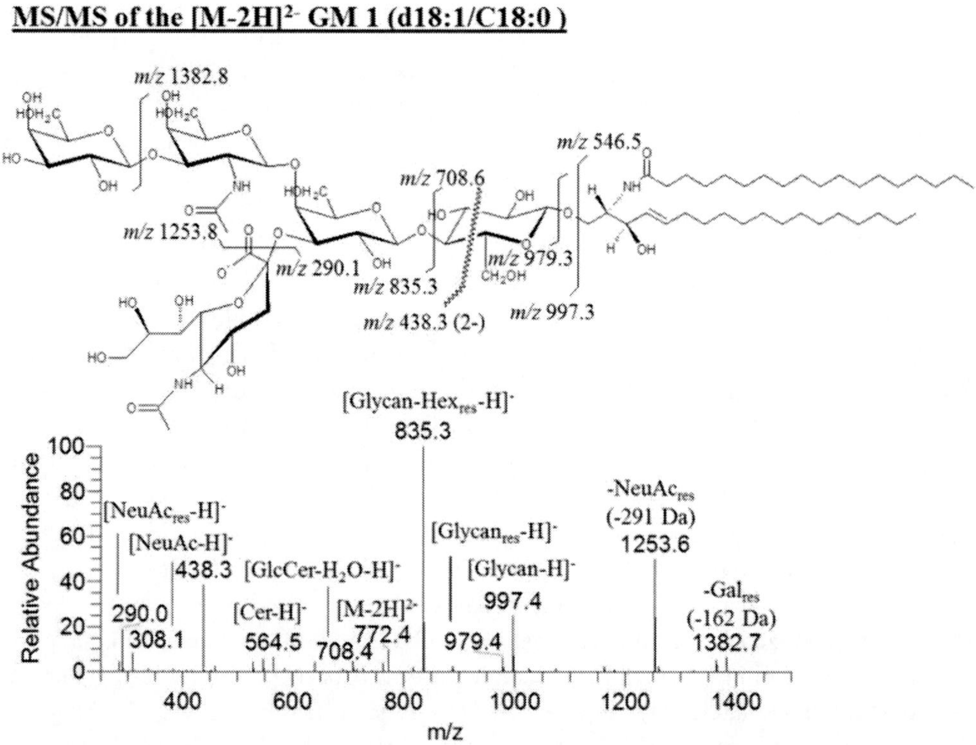

Figure 5. ESI-linear ion trap tandem mass spectra of the $[M-2H]^{2-}$ molecular ion of GM1(d18:1/C18:0) at m/z 772.

4. Sphingolipidomics in Cells, Tissues and Biological Fluids

Studies using sphingolipidomics have focused on the screening of the typical profile sphingolipids in cell, tissue and biological fluids to elucidate the biological functions of GSLs. Various approaches have been adopted, but by far the most popular is the use of LC-MRM for the identification and quantification of the molecular species of SLs extracted from the organic phase. For example, using RP C8-LC-MRM, Bui et al. profiled the sphingolipids in extracts of human plasma, and identified and quantified ceramides, hexosylceramides and glycolipids [25]. Hammad et al. measured several sphingolipids in plasma by C8-LC-MS/MS-MRM, including sphingosine and dihydrosphingosine, 1-phosphate derivatives (S1P and dhS1P), ceramides, sphingomyelins, hexosylceramides, lactosylceramides, and Cer 1-phosphate (Cer1P) [61]. These GSLs were quantified in the plasma of human volunteers in non-fasting and fasting conditions. A "normal" human serum profile of sphingolipids in plasma was established and a reference range for the circulation of the sphingolipids species was determined [61]. The species C16-SM increased with fasting, while levels of C16/C20 Cer decreased under the same conditions. Several authors have reported the quantification of ceramide species in various biological samples using LC-MS [62–65]. Most studies were performed using LC-MS combined with MRM tandem mass spectrometry with positive mode electrospray ionization: C8 LC-MS-MRM positive mode was used for the simultaneous quantitative measurement of ceramides, in human plasma, rat liver, heart and muscle tissue

[62]; C18 LC-MRM-MS was used for the identification and quantification of Cer (22:0) and Cer (24:0) in human plasma [65]; C8 Nano-LC-MS/MS analysis was used to quantify CSF ceramides in mice [66]. C18 LC-MS was used to characterize the sphingolipid composition in synovial fluid (SF) of knee joints of patients with early and late stages of osteoarthritis and rheumatoid arthritis [67]; C18 LC-MS and MALDI-MS in the negative ion mode were used to quantify sulfatides in tissues, body fluids and cells [68]; C18 LC –MS, using selected reaction monitoring (SRM) for $[M-H]^-$ and the product ion at *m/z* 97, was applied to quantify lipid rafts and non-raft membrane of the central nervous system [33].

Various LC-MS protocols have been used to screen gangliosides, extracted from the aqueous phase, from a variety of biological sources. Targeted RP ultra-performance liquid chromatography (UPLC)-MS/MS MRM was used to measure four common gangliosides (GM2, GM3, GD2, GD3) in human plasma [69]. Also, gangliosides from fetal neocortex were profiled using high resolution MS to evaluate the normal fetal neocortex lipidome in the early stages of development [32]. Normal phase LC-MS using an amino column was used for detection of GM1–3, $GD1_{a,b}$, GD2–3, and $GT1_{a,b}$ levels in serum of patients with various diseases (esophagus, adenocarcinoma, high-grade dysplasia, and gangliosidoses) [70]. GM2 levels were also detected using LC-MS in human patient samples, such as plasma, leukocytes, and cultured skin fibroblasts [71]. High resolution LC-MS mass spectrometry (LC / ESI-HR-MS) has been optimized for quantitative analysis of GD3 and GM3 in human milk [72].

Sphingolipid profiling in biological samples has also been performed using thin layer chromatography based analytical strategies. Seng et al. identified 11 classes of sphingolipids in the human lens lipidome using TLC-desorption ionization/MS [73]. This study has shown that dihydrosphingomyelin is a dominant class of sphingolipids in the human lens, and that changes in the lipidome occur during aging and cataract formation. The ganglioside profile of human bone marrow mesenchymal stem cells (hBMSCs) has also been studied by high performance TLC followed by characterization using MALDI MS. This study has demonstrated that GD1a was highly expressed by a subpopulation of hBMSCs having a marked propensity to osteogenic determination and differentiation [74]. TLC coupled to IR-MALDI-Q-TOF mass spectrometry was used to study GSLs in human hepatocellular and pancreatic tumors [19, 75]. TLC-IR-MS allowed identification of the neutral GSL profile in lipid extracts obtained from THP-1 cell [19]. Coupling TLC with ESI-MS has been used to identify not only gangliosides and STs, but also SMs, neutral GSLs, and phospholipids in extracts from bovine brain [38].

5. Sphingolipidomic Applications in Disease

a. Diabetes

The role of sphingolipids in diabetes and insulin resistance has been discussed in the literature [76–79]. Given the diversity of lipid metabolites present in biological fluids, covering a wide range of concentrations [80], their detection requires sensitive analytical strategies, such as those using mass spectrometry.

It has been shown that obese individuals with type 2 diabetes mellitus (T2DM) have increased levels of plasma ceramides, and that these correlate with the severity of the resistance to insulin [76]. Wei and colleagues observed significantly elevated levels of 1-deoxy-sphingolipids in T2DM, but not in controls or in type 1 diabetes mellitus (T1DM) patients, suggesting that these SLs could be T2DM markers [77]. These findings were recently corroborated in a cohort study of 339 individuals, which included those with metabolic syndrome (MeS) [81]. Remarkably, no changes were observed in levels of dihydroceramide, ceramides mono-, di-, tri-, GM3 and SM in blood of women with T2DM in comparison with non-diabetics [78]. Shui and colleagues reported the similarity of human lipidome and cynomolgus monkey lipidome, and observed significantly lower levels of plasma GlcCer, gangliosides GM3 and major SM species in diabetic monkeys compared to normal controls, although this was not the case with free ceramide species [79].

The amount of information on the lipidome of diabetic individuals is extensive, and provides a solid platform for the comparison and evaluation of the role of SLs in this disease. This is essential to identify potential markers and predictors of the course of the disease, for adjusting therapeutic treatments, and, ultimately, to improve the quality of life of patients. Based on literature findings, monitoring 1-deoxy-sphingolipids is emerging as a potential and reliable marker of T2DM, which might be related to the disease progression.

b. Cardiovascular Diseases

Dyslipidemia, including high cholesterol and triglyceride plasma levels, have been demonstrated to be risk factors for cardiovascular diseases (CVD). Individuals with T2DM, metabolic syndrome, obesity, kidney disease, cancer and autoimmune diseases also have increased risk of developing cardiovascular events [82–85]. Given the asymptomatic character of CVD, and the inflammatory status associated in diabetes, obesity, metabolic syndrome, auto immune diseases and other diseases, it is extremely difficult to extract the relevant lipidomic fingerprinting typical of each individual pathology. At present, cardiovascular research primarily focuses on the role of lipid classes rather than specific molecular entities [86]. Moreover, many studies have focused on the screening of abundant lipids [87, 88] with reports on the altered levels of major lipid classes including ceramides and sphingomyelins [89, 90]. However, the screening of less abundant sphingolipids is of major importance for the understanding of molecular mechanisms underlying disease, and to improve diagnosis and prognosis of cardiovascular events.

Using an LC-MS approach, it was found that C18 sphingolipids levels in plasma were significantly lower in patients with coronary artery disease, while C20 sphingolipids levels were increased [91]. Lower plasma levels of sphingolipids were observed after administration of rosuvastatin, a statin used to prevent cardiovascular events associated with dyslipidemia [92]. In this study, the authors reported a significant dose-dependent reduction with rosuvastatin in ceramides, sphingomyelins, monohexosylceramides, dihexosylceramides, trihexosylceramides, and GM3 gangliosides, while LDL levels had no influence on these SLs [92].

c. Cancer

It is well established that specific glycosylation patterns, including those of glycosylated lipids, such as glycosphingolipids, have biological significance in the different stages of tumor development [93]. Although a series of publications have reviewed the role of sphingolipids as modulators of apoptosis, differentiation, growth, adhesion, survival of cancer cells (discussed in [94, 95]), little effort has been devoted to screening SLs in cancer cells lines. Some studies have indicated changes in sphingomyelin and ceramide levels associated with malignancy [35, 96–98]. Also, increased levels of SM were reported in non-metastatic and metastatic breast cancer cells [35] and in plasma of patients with prostate cancer [97]. However, in a cohort study of 184 patients, a decrease of SM content was observed in malignant lung tissue [100].

In these studies, less abundant and minority SL species, which are present in less than <10% in biological fluids [61, 99], are usually overlooked and remain largely unaccounted for. These include hexosylceramides (HexCer), lactosylceramides (LacCer), sulfatides (ST), sphingosine-1-phosphate (S1P), gangliosides (Gan), ceramide-1-phosphate (Cer1P), and di-hydroceramides (dhCer) (figure 1). Until now, the number of untargeted metabolic studies able to provide a comprehensive description of the sphingolipids in cancer is very scarce [100], and this area requires further studies.

d. Auto Immune Diseases

Auto immune diseases (AID) constitute a wide variety of pathological conditions (over 100), which involve the mucous, gastric, cutaneous, neuronal, connective, lung, bone and other tissues with no clearly defined symptoms, and affect 1-2% of the world population in developed countries [101]. Although their relevance is well recognized, there is a need of more specific disease markers, other than the common ones associated with the inflammatory status (increased CRP, IL-6 and TNF-alpha) that are characteristic during the initial stages (outbreak).

Recently, McDonald and colleagues reported the accumulation of glycosphingolipids (lactosylceramides and gangliosides GM3, GA2, Gb3, Gb4 and GM2) in CD4+ T cells of patients with systemic lupus erythematosus (SLE) [102]. They suggested that altered GSL levels in leukocytes were associated with defects in biosynthesis, transport or recycling pathways, as inhibition of GSL biosynthesis led to the normalization GSL levels and restored leukocyte function, attenuating the autoimmunity in SLE patients [102]. These findings are not consistent with those described by other authors, which have reported changes to GM1 and sulfatide in SLE patients and Sjogren's syndrome [103]. It is known that GSLs are located at the surface of membranes and are involved in activation of T cell subtypes [104]. However, the potential of GSLs as markers of AID remains to be corroborated by additional studies.

So far, the use of large-scale lipidomics in AID is scarce [105–107]. Elevated levels of galactosylceramides and GM1 gangliosides were noticed in patients with multiple sclerosis [108]. Other targeted lipidomic studies were performed in patients with Crohn's disease [106], arthritis [98, 109], systemic lupus erythematosus [107] and cystic fibrosis [95]. Studies have shown altered levels of fatty acids in rheumatoid arthritis [96], and of lysoPC/PC ratios in patients with Multiple Sclerosis [105], and deregulation of arachidonic-containing lipids in osteoarthritis [109]. There is some evidence that different GSLs are altered, depending on the nature of the AID [102], but no characteristic sphingolipid signatures for AID have been found.

The wide array of AID conditions, the large range of fluids and tissues studied [106,108–110] and the variety of analytical strategies used are limiting factors to obtaining complete knowledge of the composition and role of sphingolipids in AID. Moreover, the basal values or the variation introduced by other factors such as gender, age, ethnicity, eating, drinking, smoking habits, and lipid lowering drugs have not yet been considered.

6. Challenges and Future Directions

Because of the large diversity of sphingolipid structures present over a wide range of concentrations in biological fluids and tissues, the actual sphingolipidomic approach used in most studies is not very effective for the systematic analysis of the sphingolipidome. There is a need to develop new approaches that will allow us to obtain detailed structural and quantitative information, of all the sphingolipid species with high sensitivity. Nevertheless, using the available lipidomic approaches it has been possible to screen sphingolipids in biological systems and to provide important information on the sphingolipidome in chronic diseases such as diabetes, cancer, cardiovascular and autoimmune diseases.

Many technological and biochemical challenges remain before successful implementation of technologies aimed at the full characterization of the sphingolipidome in health and disease. This will enable us to unveil the pathways and networks of cellular sphingolipids in biological systems, and, ultimately, to discover SLs as potential biomarkers of disease states, and to identify potential novel drug targets.

Acknowledgments

Thanks are due to Fundação para a Ciência e a Tecnologia (FCT, Portugal), European Union, QREN, FEDER, and COMPETE for funding the QOPNA research unit (project PEst-C/QUI/UI0062/2013; FCOMP-01-0124-FEDER-037296), and RNEM (REDE/1504/REM /2005 that involves the Portuguese National Mass Spectrometry Network). AR greatly acknowledges the financial support provided by the Fundação para a Ciência e Tecnologia (SFRH/BPD/101916/2014).

References

[1] Hong, JY; Allegood, JC; Kelly, S; Wang, E; Merrill, AH; Wang, MD. *Sphingolipid Biology*, Springer Japan, Tokyo, 2006.

[2] Regina Todeschini, A; Hakomori, SI. Functional role of glycosphingolipids and gangliosides in control of cell adhesion, motility, and growth, through glycosynaptic microdomains, *Biochim. Biophys. Acta - Gen. Subj.*, 1780 (2008), 421–433.

[3] Bartke, N; Hannun, Ya. Bioactive sphingolipids: metabolism and function., *J. Lipid Res.*, 50 Suppl (2009), S91–S96.

[4] Pruett, ST; Bushnev, A; Hagedorn, K; Adiga, M; Haynes, CA; Sullards, MC; Liotta, DC; Merrill, AH. Biodiversity of sphingoid bases ("sphingosines") and related amino alcohols., *J. Lipid Res.*, 49 (2008), 1621–39.

[5] Chester, MA. Nomenclature of Glycolipids (IUPAC Recommendations 1997), *Pure Appl. Chem.*, 69 (1997), 2475–2488.

[6] Nelson, DL; Lehninger, AL; Cox, MM. *Lehninger Principles of Biochemistry*, W. H. Freeman, 2008.

[7] Posse de Chaves, E; Sipione, S. Sphingolipids and gangliosides of the nervous system in membrane function and dysfunction, *FEBS Lett.*, 584 (2010), 1748–1759.

[8] Furukawa, K; Ohmi, Y; Ohkawa, Y; Tokuda, N; Kondo, Y; Tajima, O; Furukawa, K. Regulatory mechanisms of nervous systems with glycosphingolipids, *Neurochem. Res.*, 36 (2011), 1578–1586.

[9] Haughey, NJ. Sphingolipids in neurodegeneration, *NeuroMolecular Med.*, 12 (2010), 301–305.

[10] Patwardhan, Ga; Liu, Y-Y. Sphingolipids and expression regulation of genes in cancer., *Prog. Lipid Res.*, 50 (2011), 104–114.

[11] Duan, RD; Nilsson, Å. Metabolism of sphingolipids in the gut and its relation to inflammation and cancer development, *Prog. Lipid Res.*, 48 (2009), 62–72.

[12] Xu, Y-H; Barnes, S; Sun, Y; Grabowski, GA. Multi-system disorders of glycosphingolipid and ganglioside metabolism., *J. Lipid Res.*, 51 (2010), 1643–75.

[13] Platt, FM. Sphingolipid lysosomal storage disorders., *Nature.*, 510 (2014), 68–75.

[14] Merrill, AH. Sphingolipid and glycosphingolipid metabolic pathways in the era of sphingolipidomics, *Chem. Rev.*, 111 (2011) 6387–6422.

[15] Haynes, CA; Allegood, JC; Park, H; Sullards, MC. Sphingolipidomics: methods for the comprehensive analysis of sphingolipids., J. Chromatogr. B. Analyt. *Technol. Biomed. Life Sci.*, 877 (2009), 2696–708.

[16] Meisen, I; Mormann, M; Müthing, J. Thin-layer chromatography, overlay technique and mass spectrometry: A versatile triad advancing glycosphingolipidomics, *Biochim. Biophys. Acta - Mol. Cell Biol. Lipids.*, 1811 (2011), 875–896.

[17] Fuchs, B. Analysis of phospolipids and glycolipids by thin-layer chromatography-matrix-assisted laser desorption and ionization mass spectrometry, *J. Chromatogr. A.*, 1259 (2012), 62–73.

[18] Fuchs, B; Süß, R; Nimptsch, A; Schiller, J. MALDI-TOF-MS Directly Combined with TLC: *A Review of the Current State*, *Chromatographia.*, 69 (2009), 95–105.

[19] Kouzel, IU; Pirkl, A; Pohlentz, G; Soltwisch, J; Dreisewerd, K; Karch, H; Müthing, J. Progress in detection and structural characterization of glycosphingolipids in crude lipid extracts by enzymatic phospholipid disintegration combined with thin-layer chromatography immunodetection and IR-MALDI mass spectrometry., *Anal. Chem.*, 86 (2014), 1215–22.

[20] Merrill, AH; Sullards, MC; Allegood, JC; Kelly, S; Wang, E. Sphingolipidomics: high-throughput, structure-specific, and quantitative analysis of sphingolipids by liquid chromatography tandem mass spectrometry., *Methods.*, 36 (2005), 207–24.

[21] Farwanah, H; Kolter, T. *Lipidomics of Glycosphingolipids*, (2012), 134–164.

[22] Folch, J; Lees, M; Sloane Stanley, GH. A simple method for the isolation and purification of total lipides from animal tissues., *J. Biol. Chem.*, 226 (1957), 497–509.

[23] Bligh, EG; Dyer, WJ. A rapid method of total lipid extraction and purification., *Can. J. Biochem. Physiol.*, 37 (1959), 911–917.

[24] Matyash, V; Liebisch, G; Kurzchalia, TV; Shevchenko, A; Schwudke, D. Lipid extraction by methyl-tert-butyl ether for high-throughput lipidomics., *J. Lipid Res.*, 49 (2008), 1137–1146.

[25] Bui, HH; Leohr, JK; Kuo, M-S. Analysis of sphingolipids in extracted human plasma using liquid chromatography electrospray ionization tandem mass spectrometry, *Anal. Biochem.*, 423 (2012), 187–194.

[26] Liebisch, G; Scherer, M. Quantification of bioactive sphingo- and glycerophospholipid species by electrospray ionization tandem mass spectrometry in blood, *J. Chromatogr. B Anal. Technol. Biomed. Life Sci.*, 883-884 (2012), 141–146.

[27] Masood, MA; Rao, RP; Acharya, JK; Blonder, J; Veenstra, TD. Quantitation of multiple sphingolipid classes using normal and reversed-phase LC-ESI-MS/MS: comparative profiling of two cell lines., *Lipids.*, 47 (2012), 209–26.

[28] Han, X; Jiang, X. A review of lipidomic technologies applicable to sphingolipidomics and their relevant applications., *Eur. J. Lipid Sci. Technol.*, 111 (2009), 39–52.

[29] Lingwood, Ca. Glycosphingolipid functions, Cold Spring Harb. *Perspect. Biol.*, 3 (2011), 1–26.

[30] Torretta, E; Vasso, M; Fania, C; Capitanio, D; Bergante, S; Piccoli, M; Tettamanti, G; Anastasia, L; Gelfi, C. Application of direct HPTLC-MALDI for the qualitative and quantitative profiling of neutral and acidic glycosphingolipids: The case of NEU3 overexpressing C2C12 murine myoblasts, *Electrophoresis.*, (2014), 1319–1328.

[31] Müthing, J; Distler, U. Advances on the compositional analysis of glycosphingolipids combining thin-layer chromatography with mass spectrometry., *Mass Spectrom. Rev.*, 29 (n.d.), 425–79.

[32] Ghiulai, RM; Sarbu, M; Vukelić, Ž; Ilie, C; Zamfir, AD. Early stage fetal neocortex exhibits a complex ganglioside profile as revealed by high resolution tandem mass spectrometry, *Glycoconj. J.*, 31 (2014), 231–245.

[33] Moyano, AL; Li, G; Lopez-Rosas, A; Månsson, J-E; van Breemen, RB; Givogri, MI. Distribution of C16:0, C18:0, C24:1, and C24:0 sulfatides in central nervous system lipid rafts by quantitative ultra-high-pressure liquid chromatography tandem mass spectrometry, *Anal. Biochem.*, 467 (2014), 31–39.

[34] Melo, T; Videira, RA; André, S; Maciel, E; Francisco, CS; Oliveira-Campos, AM; Rodrigues, LM; Domingues, MRM; Peixoto, F; Oliveira, MM. Tacrine and its analogues impair mitochondrial function and bioenergetics: A lipidomic analysis in rat brain, *J. Neurochem.*, 120 (2012), 998–1013.

[35] Dória, ML; Cotrim, Z; Macedo, B; Simões, C; Domingues, P; Helguero, L; Domingues, MR; Lipidomic approach to identify patterns in phospholipid profiles and define class differences in mammary epithelial and breast cancer cells, *Breast Cancer Res. Treat.*, 133 (2012), 635–648.

[36] Santinha, DR; Doria, ML; Neves, BM; Maciel, EA; Martins, J; Helguero, L; Domingues, P; Cruz, MT; Domingues, MR. Prospective phospholipid markers for skin sensitization prediction in keratinocytes: a phospholipidomic approach, *Arch Biochem Biophys.*, 533 (2013), 33–41.

[37] Ivleva, VB; Sapp, LM; O'Connor, PB; Costello, CE. Ganglioside analysis by thin-layer chromatography matrix-assisted laser desorption/ionization orthogonal time-of-flight mass spectrometry, *J. Am. Soc. Mass Spectrom.*, 16 (2005), 1552–1560.

[38] Park, H; Zhou, Y; Costello, CE. Direct analysis of sialylated or sulfated glycosphingolipids and other polar and neutral lipids using TLC-MS interfaces., *J. Lipid Res.*, 55 (2014), 773–81.

[39] Valdes-Gonzalez, T; Goto-Inoue, N; Hirano, W; Ishiyama, H; Hayasaka, T; Setou, M; Taki, T. New approach for glyco- and lipidomics - Molecular scanning of human brain gangliosides by TLC-Blot and MALDI-QIT-TOF MS, *J. Neurochem.*, 116 (2011), 678–683.

[40] Taki, T. Frontier in Tumor Glycobiology An Approach to Glycobiology from Glycolipidomics: Ganglioside Molecular Scanning in the Brains of Patients with Alzheimer' s Disease by TLC-Blot / Matrix Assisted Laser Desorption/Ionization-Time of Flight MS, *Bio. Pharm. Bull.*, 35 (2012), 1642–1647.

[41] Arigi, E; Blixt, O; Buschard, K; Clausen, H; Levery, SB. Design of a covalently bonded glycosphingolipid microarray, *Glycoconj. J.*, 29 (2012), 1–12.

[42] Sakumoto, Y; Ueta, H; Yuki, N; Matsuno, K. Simultaneous immunohistochemical detection of gangliosides and neuronal markers in paraformaldehyde-fixed nervous tissues by acetone etching., *Arch. Histol. Cytol.*, 72 (2009), 77–90.

[43] Kuchar, L; Faltyskova, H; Krasny, L; Dobrovolny, R; Hulkova, H; Ledvinova, J; Volny, M; Strohalm, M; Lemr, K; Kryspinova, L; Asfaw, B; Rybová, J; Desnick, RJ; Havlicek, V. Fabry disease: renal sphingolipid distribution in the α-Gal A knockout mouse model by mass spectrometric and immunohistochemical imaging, *Anal. Bioanal. Chem.*, 407 (2014), 2283–2291.

[44] Farwanah, H; Kolter, T; Sandhoff, K. Mass spectrometric analysis of neutral sphingolipids: Methods, applications, and limitations, Biochim. Biophys. *Acta - Mol. Cell Biol. Lipids.*, 1811 (2011), 854–860.

[45] Hsu, F-F; Turk, J. Studies on sulfatides by quadrupole ion-trap mass spectrometry with electrospray ionization: structural characterization and the fragmentation processes that include an unusual internal galactose residue loss and the classical charge-remote fragmentation., *J. Am. Soc. Mass Spectrom.*, 15 (2004), 536–46.

[46] Ito, E; Tominaga, A; Waki, H; Miseki, K; Tomioka, A; Nakajima, K; Kakehi, K; Suzuki, M; Taniguchi, N; Suzuki, A. Structural characterization of monosialo-, disialo- and trisialo-gangliosides by negative ion AP-MALDI-QIT-TOF mass spectrometry with MS(n) switching., *Neurochem. Res.*, 37 (2012), 1315–24.

[47] Lee, H; Lerno, LA; Choe, Y; Chu, CS; Gillies, LA; Grimm, R; Lebrilla, CB; German, JB. Multiple precursor ion scanning of gangliosides and sulfatides with a reversed-phase microfluidic chip and quadrupole time-of-flight mass spectrometry., *Anal. Chem.*, 84 (2012), 5905–12.

[48] Ii, T; Ohashi, Y; Nagai, Y. Structural elucidation of underivatized gangliosides by electrospray-ionization tandem mass spectrometry (ESIMS/MS)., *Carbohydr. Res.*, 273 (1995), 27–40.

[49] Chen, Y; Liu, Y; Sullards, MC; Merrill, AH. An introduction to sphingolipid metabolism and analysis by new technologies., *Neuromolecular Med.*, 12 (2010), 306–19.

[50] Brugger, B; Erben, G; Sandhoff, R; Wieland, FT; Lehmann, WD. Quantitative analysis of biological membrane lipids at the low picomole level by nano-electrospray ionization tandem mass spectrometry, *Proc. Natl. Acad. Sci.*, 94 (1997), 2339–2344.

[51] Pulfer, M; Murphy, R. Electrospray mass spectrometry of phospholipids, *Mass Spectrom. Rev.*, (2003).

[52] Gu, M; Kerwin, JL; Watts, JD; Aebersold, R. Ceramide profiling of complex lipid mixtures by electrospray ionization mass spectrometry., *Anal. Biochem.*, 244 (1997), 347–56.

[53] Ann, Q; Adams, J. Structure determination of ceramides and neutral glycosphingolipids by collisional activation of [M + Li](+) ions., *J. Am. Soc. Mass Spectrom.*, 3 (1992), 260–3.

[54] Ann, Q; Adams, J. Structure-specific collision-induced fragmentations of ceramides cationized with alkali-metal ions, *Anal. Chem.*, 65 (1993), 7–13.

[55] Cameron Sullards, M. Sphingolipid Metabolism and Cell Signaling, Part B, Elsevier, 2000.

[56] Jiang, X; Han, X. Characterization and direct quantitation of sphingoid base-1-phosphates from lipid extracts: a shotgun lipidomics approach., *J. Lipid Res.*, 47 (2006), 1865–73.

[57] Hsu, F-F; Turk, J. Structural determination of glycosphingolipids as lithiated adducts by electrospray ionization mass spectrometry using low-energy collisional-activated dissociation on a triple stage quadrupole instrument, *J. Am. Soc. Mass Spectrom.*, 12 (2001), 61–79.

[58] Domon, B; Costello, CE. Structure elucidation of glycosphingolipids and gangliosides using high-performance tandem mass spectrometry, *Biochemistry.*, 27 (1988), 1534–1543.

[59] Ohashi, Y; Nagai, Y. Fast-atom-bombardment chemistry of sulfatide (3-sulfogalactosylceramide), *Carbohydr. Res.*, 221 (1991), 235–243.

[60] Juhasz, P; Costello, CE. Matrix-assisted laser desorption ionization time-of-flight mass spectrometry of underivatized and permethylated gangliosides., *J. Am. Soc. Mass Spectrom.*, 3 (2006), 785–96.

[61] Hammad, SM; Pierce, JS; Soodavar, F; Smith, KJ; Al Gadban, MM; Rembiesa, B; Klein, RL; Hannun, Ya; Bielawski, J; Bielawska, A. Blood sphingolipidomics in healthy humans: impact of sample collection methodology., *J. Lipid Res.*, 51 (2010), 3074–3087.

[62] Kasumov, T; Huang, H; Chung, YM; Zhang, R; McCullough, AJ; Kirwan, JP. Quantification of ceramide species in biological samples by liquid chromatography electrospray ionization tandem mass spectrometry, *Anal. Biochem.*, 401 (2010), 154–161.

[63] Haynes, TAS; Duerksen-Hughes, PJ; Filippova, M; Filippov, V; Zhang, K. C18 ceramide analysis in mammalian cells employing reversed-phase high-performance liquid chromatography tandem mass spectrometry., *Anal. Biochem.*, 378 (2008), 80–86.

[64] Wang, JR; Zhang, H; Yau, LF; Mi, JN; Lee, S; Lee, KC; Hu, P; Liu, L; Jiang, ZH. Improved sphingolipidomic approach based on ultra-high performance liquid chromatography and multiple mass spectrometries with application to cellular neurotoxicity, *Anal. Chem.*, 86 (2014), 5688–5696.

[65] Jiang, H; Hsu, FF; Farmer, MS; Peterson, LR; Schaffer, JE; Ory, DS; Jiang, X. Development and validation of LC-MS/MS method for determination of very long acyl chain (C22:0 and C24:0) ceramides in human plasma, *Anal. Bioanal. Chem.*, 405 (2013), 7357–7365.

[66] Thomas, D; Eberle, M; Schiffmann, S; Zhang, DD; Geisslinger, G; Ferreirós, N. Nano-LC-MS/MS for the quantitation of ceramides in mice cerebrospinal fluid using minimal sample volume, *Talanta.*, 116 (2013), 912–918.

[67] Kosinska, MK; Liebisch, G; Lochnit, G; Wilhelm, J; Klein, H; Kaesser, U; Lasczkowski, G; Rickert, M; Schmitz, G; Steinmeyer, J. Sphingolipids in human synovial fluid - A lipidomic study, *PLoS One.*, 9 (2014).

[68] Mirzaian, M; Kramer, G; Poorthuis, BJHM. Quantification of sulfatides and lysosulfatides in tissues and body fluids by liquid chromatography-tandem mass spectrometry, *J. Lipid Res.*, 56 (2015), 936–943.

[69] Huang, Q; Zhou, X; Liu, D; Xin, B; Cechner, K; Wang, H; Zhou, A. A new liquid chromatography/tandem mass spectrometry method for quantification of gangliosides in human plasma., *Anal. Biochem.*, 455 (2014), 26–34.

[70] Garcia, AD; Chavez, JL; Mechref, Y. Rapid and sensitive LC-ESI-MS of gangliosides., J. Chromatogr. B. Analyt. *Technol. Biomed. Life Sci.*, 947-948 (2014), 1–7.

[71] Fuller, M; Duplock, S; Hein, LK; Rigat, Ba; Mahuran, DJ. Liquid chromatography/electrospray ionisation-tandem mass spectrometry quantification of GM2 gangliosides in human peripheral cells and plasma, *Anal. Biochem.*, 458 (2014), 20–26.

[72] Giuffrida, F; Elmelegy, IM; Thakkar, SK; Marmet, C; Destaillats, F. Longitudinal Evolution of the Concentration of Gangliosides GM3 and GD3 in Human Milk, *Lipids.*, 49 (2014), 997–1004.

[73] Seng, JA; Ellis, SR; Hughes, JR; Maccarone, AT; Truscott, RJW; Blanksby, SJ; Mitchell, TW. Characterisation of sphingolipids in the human lens by thin layer chromatography-desorption electrospray ionisation mass spectrometry., *Biochim. Biophys. Acta.*, 1841 (2014), 1285–91.

[74] Bergante, S; Torretta, E; Creo, P; Sessarego, N; Papini, N; Piccoli, M; Fania, C; Cirillo, F; Conforti, E; Ghiroldi, A; Tringali, C; Venerando, B; Ibatici, A; Gelfi, C; Tettamanti, G; Anastasia, L. Gangliosides as a potential new class of stem cell markers: the case of GD1a in human bone marrow stem cells., *J. Lipid Res.*, 55 (2014), 549–560.

[75] Distler, U; Hülsewig, M; Souady, J; Dreisewerd, K; Haier, J; Senninger, N; Friedrich, AW; Karch, H; Hillenkamp, F; Berkenkamp, S; Peter-Katalinić, J; Müthing, J. Matching IR-MALDI-o-TOF mass spectrometry with the TLC overlay binding assay and its clinical application for tracing tumor-associated glycosphingolipids in hepatocellular and pancreatic cancer, *Anal. Chem.*, 80 (2008), 1835–1846.

[76] Haus, JM; Kashyap, SR; Kasumov, T; Zhang, R; Kelly, KR; Defronzo, RA; Kirwan, JP. Plasma ceramides are elevated in obese subjects with type 2 diabetes and correlate with the severity of insulin resistance., *Diabetes.*, 58 (2009), 337–43.

[77] Wei, N; Pan, J; Pop-busui, R; Othman, A; Alecu, I; Hornemann, T; Eichler, FS. *Altered sphingoid base profiles in type 1 compared to type 2 diabetes*, (2014), 2–5.

[78] Lappas, M; Mundra, Pa; Wong, G; Huynh, K; Jinks, D; Georgiou, HM; Permezel, M; Meikle, PJ. The prediction of type 2 diabetes in women with previous gestational diabetes mellitus using lipidomics, *Diabetologia.*, (2015).

[79] Shui, G; Stebbins, JW; Lam, BD; Cheong, WF; Lam, SM; Gregoire, F; Kusonoki, J; Wenk, MR. Comparative plasma lipidome between human and cynomolgus monkey: are plasma polar lipids good biomarkers for diabetic monkeys? *PLoS One.*, 6 (2011), e19731.

[80] Quehenberger, O; Dennis, EA. The human plasma lipidome., *N. Engl. J. Med.*, 365 (2011), 1812–23.

[81] Othman, a; Saely, CH; Muendlein, a; Vonbank, a; Drexel, H; von Eckardstein, a; Hornemann, T. Plasma 1-deoxysphingolipids are predictive biomarkers for type 2 diabetes mellitus, *BMJ Open Diabetes Res. Care.*, 3 (2015), e000073–e000073.

[82] Querton, L; Buysschaert, M; Hermans, MP. Hypertriglyceridemia and residual dyslipidemia in statin-treated, patients with diabetes at the highest risk for cardiovascular disease and achieving very-low low-density lipoprotein-cholesterol levels, *J. Clin. Lipidol.*, 6 (2012), 434–442.

[83] Ma, Y; Zhou, L; Dong, J; Zhang, X; Yan, S. Arterial stiffness and increased cardiovascular risk in chronic kidney disease, *Int. Urol. Nephrol.*, (2015).

[84] Chen, SC; Hung, CC; Kuo, MC; Lee, JJ; Chiu, YW; Chang, JM; Hwang, SJ; Chen, HC; Association of Dyslipidemia with Renal Outcomes in Chronic Kidney Disease, *PLoS One.*, 8 (2013).

[85] Georgiadis, AN; Papavasiliou, EC; Lourida, ES; Alamanos, Y; Kostara, C; Tselepis, AD; Drosos, Aa. Atherogenic lipid profile is a feature characteristic of patients with early rheumatoid arthritis: effect of early treatment--a prospective, controlled study., *Arthritis Res. Ther.*, 8 (2006), R82.

[86] Hinterwirth, H; Stegemann, C; Mayr, M. Lipidomics: Quest for Molecular Lipid Biomarkers in Cardiovascular Disease, *Circ. Cardiovasc. Genet.*, 7 (2014), 941–954.

[87] Stübiger, G; Aldover-Macasaet, E; Bicker, W; Sobal, G; Willfort-Ehringer, A; Pock, K; Bochkov, V; Widhalm, K; Belgacem, O. Targeted profiling of atherogenic phospholipids in human plasma and lipoproteins of hyperlipidemic patients using MALDI-QIT-TOF-MS/MS, *Atherosclerosis.*, 224 (2012), 177–186.

[88] Bojic, L; McLaren, D; Shah, V; Previs, S; Johns, D; Castro-Perez, J. Lipidome of Atherosclerotic Plaques from Hypercholesterolemic Rabbits, *Int. J. Mol. Sci.*, 15 (2014), 23283–23293.

[89] Ichi, I; Nakahara, K; Miyashita, Y; Hidaka, A; Kutsukake, S; Inoue, K; Maruyama, T; Miwa, Y; Harada-Shiba, M; Tsushima, M; Kojo, S. Association of ceramides in human plasma with risk factors of atherosclerosis, *Lipids.*, 41 (2006), 859–863.

[90] Yu, J; Pan, W; Shi, R; Yang, T; Li, Y; Yu, G; Bai, Y; Schuchman, EH; He, X; Zhang, G. Ceramide Is Upregulated and Associated With Mortality in Patients With Chronic Heart Failure, *Can. J. Cardiol.*, 31 (2015), 357–363.

[91] Othman, A; Saely, CH; Muendlein, A; Vonbank, A; Drexel, H; von Eckardstein, A; Hornemann, T. Plasma C20-Sphingolipids predict cardiovascular events independently from conventional cardiovascular risk factors in patients undergoing coronary angiography, *Atherosclerosis.*, 240 (2015), 216–221.

[92] Ng, TWK; Ooi, EMM; Watts, GF; Chan, DC; Weir, JM; Meikle, PJ; Barrett, PHR. Dose-Dependent Effects of Rosuvastatin on the Plasma Sphingolipidome and Phospholipidome in the Metabolic Syndrome, *J. Clin. Endocrinol. Metab.*, 99 (2014), E2335–E2340.

[93] Hakomori, S. Tumor Malignancy Defined by Aberrant Glycosylation and Sphingo (glyco) lipid Metabolism Tumor Malignancy Defined by Aberrant Glycosylation and Sphingo (glyco) lipid Metabolism,' *Cancer Res*., (1996), 5309–5318.

[94] Giussani, P; Tringali, C; Riboni, L; Viani, P; Venerando, B. Sphingolipids: Key regulators of apoptosis and pivotal players in cancer drug resistance, *Int. J. Mol. Sci.*, 15 (2014), 4356–4392.

[95] Ponnusamy, S; Meyers-Needham, M; Senkal, CE; Saddoughi, SA; Sentelle, D; Selvam, SP; A. Salas, B. Ogretmen, Sphingolipids and cancer: ceramide and sphingosine-1-phosphate in the regulation of cell death and drug resistance., *Future Oncol*., 6 (2010), 1603–24.

[96] Fillet, M; Van Heugen, JC; Servais, aC; De Graeve, J; Crommen, J. Separation, identification and quantitation of ceramides in human cancer cells by liquid chromatography-electrospray ionization tandem mass spectrometry, *J. Chromatogr. A*., 949 (2002), 225–233.

[97] Zhou, X; Mao, J; Ai, J; Deng, Y; Roth, MR; Pound, C; Henegar, J; Welti, R; Bigler, Sa. Identification of Plasma Lipid Biomarkers for Prostate Cancer by Lipidomics and Bioinformatics, *PLoS One*., 7 (2012).

[98] Marien, E; Meister, M; Muley, T; Fieuws, S; Bordel, S; Derua, R; Spraggins, J; Van de Plas, R; Dehairs, J; Wouters, J; Bagadi, M; Dienemann, H; Thomas, M; Schnabel, Pa; Caprioli, RM; Waelkens, E; Swinnen, JV. Non-small cell lung cancer is characterized by dramatic changes in phospholipid profiles, *Int. J. Cancer*., 00 (2015) n/a–n/a.

[99] Scherer, M; Böttcher, A; Schmitz, G; Liebisch, G. Sphingolipid profiling of human plasma and FPLC-separated lipoprotein fractions by hydrophilic interaction chromatography tandem mass spectrometry., *Biochim. Biophys. Acta*., 1811 (2011), 68–75.

[100] Valsecchi, M; Aureli, M; Mauri, L; Illuzzi, G; Chigorno, V; Prinetti, A; Sonnino, S. Sphingolipidomics of A2780 human ovarian carcinoma cells treated with synthetic retinoids., *J. Lipid Res*., 51 (2010), 1832–1840.

[101] Cooper, GS; Bynum, MLK; Somers, EC. Recent insights in the epidemiology of autoimmune diseases: improved prevalence estimates and understanding of clustering of diseases., *J. Autoimmun*., 33 (2009), 197–207.

[102] McDonald, G; Deepak, S; Miguel, L; Hall, CJ; Isenberg, Da; Magee, AI; Butters, T; Jury, EC. Normalizing glycosphingolipids restores function in CD4+ T cells from lupus patients, *J. Clin. Invest*., 124 (2014), 712–724.

[103] Alpa, M; Ferrero, B; Cavallo, R; Perna, a; Naretto, C; Gennaro, M; Di Simone, D; Bellizia, L; Mansouri, M; Rossi, D; Modena, V; Giachino, O; Sena, LM; Roccatello, D. Anti-GM1 and anti-sulfatide antibodies in patients with systemic lupus erythematosus, Sjögren's syndrome, mixed cryoglobulinemia and idiopathic systemic vasculitis., *Clin. Exp. Rheumatol*., 25 (2007), 556–562.

[104] Inokuchi, J; Nagafuku, M; Ohno, I; Suzuki, A. Distinct selectivity of gangliosides required for CD4+ T and CD8+ T cell activation, Biochim. Biophys. *Acta - Mol. Cell Biol. Lipids*., 1851 (2015), 98–106.

[105] Del Boccio, P; Pieragostino, D; Di Ioia, M; Petrucci, F; Lugaresi, A; De Luca, G; Gambi, D; Onofrj, M; Di Ilio, C; Sacchetta, P; Urbani, A. Lipidomic investigations for the characterization of circulating serum lipids in multiple sclerosis, *J. Proteomics*., 74 (2011), 2826–2836.

[106] Sewell, GW; Hannun, Ya; Han, X; Koster, G; Bielawski, J; Goss, V; Smith, PJ; Rahman, FZ; Vega, R; Bloom, SL; Walker, AP; Postle, AD; Segal, AW. Lipidomic profiling in Crohn's disease: Abnormalities in phosphatidylinositols, with preservation of ceramide, phosphatidylcholine and phosphatidylserine composition, *Int. J. Biochem. Cell Biol.*, 44 (2012), 1839–1846.

[107] Jovanović, V; Abdul Aziz, N; Lim, YT; Ng Ai Poh, A; Jin Hui Chan, S; Ho Xin Pei, E; Lew, FC; Shui, G; Jenner, AM; Bowen, L; McKinney, EF; Lyons, Pa; Kemeny, MD; Smith, KGC; Wenk, MR; MacAry, Pa. Lipid Anti-Lipid Antibody Responses Correlate with Disease Activity in Systemic Lupus Erythematosus, *PLoS One.*, 8 (2013).

[108] Haghighi, S; Lekman, a; Nilsson, S; Blomqvist, M; Andersen, O. Increased CSF sulfatide levels and serum glycosphingolipid antibody levels in healthy siblings of multiple sclerosis patients, *J. Neurol. Sci.*, 326 (2013), 35–39.

[109] Castro-perez, JM; Kamphorst, J; Degroot, J; Lafeber, F; Goshawk, J; Yu, K; Shockcor, JP; Vreeken, RJ; Hankemeier, T. *Comprehensive LC - MS E Lipidomic Analysis using a Shotgun Approach and Its Application to Biomarker Detection and Identification in Osteoarthritis Patients research articles*, (2010), 2377–2389.

[110] Giera, M; Ioan-Facsinay, A; Toes, R; Gao, F; Dalli, J; Deelder, AM; Serhan, CN; Mayboroda, Oa. Lipid and lipid mediator profiling of human synovial fluid in rheumatoid arthritis patients by means of LC-MS/MS, Biochim. *Biophys. Acta - Mol. Cell Biol. Lipids.*, 1821 (2012), 1415–1424.

[111] Sullards, MC; Allegood, JC; Kelly, S; Wang, E; Haynes, CA; Park, H; Chen, Y; Merrill, AH. Structure-specific, quantitative methods for analysis of sphingolipids by liquid chromatography-tandem mass spectrometry: "inside-out" sphingolipidomics., *Methods Enzymol.*, 432 (2007), 83–115.

[112] Schwudke, D; Oegema, J; Burton, L; Entchev, E; Hannich, JT; Ejsing, CS; Kurzchalia, T; Shevchenko, A. Lipid profiling by multiple precursor and neutral loss scanning driven by the data-dependent acquisition., *Anal. Chem.*, 78 (2006), 585–95.

[113] Ekroos, K; Chernushevich, IV; Simons, K; Shevchenko, A. Quantitative Profiling of Phospholipids by Multiple Precursor Ion Scanning on a Hybrid Quadrupole Time-of-Flight Mass Spectrometer, *Anal. Chem.*, 74 (n.d.), 941–949.

[114] Houjou, T; Yamatani, K; Nakanishi, H; Imagawa, M; Shimizu, T; Taguchi, R. Rapid and selective identification of molecular species in phosphatidylcholine and sphingomyelin by conditional neutral loss scanning and MS3., Rapid Commun. *Mass Spectrom.*, 18 (2004), 3123–30.

[115] Mills, K; Johnson, A; Winchester, B. Synthesis of novel internal standards for the quantitative determination of plasma ceramide trihexoside in Fabry disease by tandem mass spectrometry, *FEBS Lett.*, 515 (2002), 171–176.

[116] Yildiz, Y; Matern, H; Thompson, B; Allegood, JC; Warren, RL; Ramirez, DMO; Hammer, RE; Hamra, FK; Matern, S; Russell, DW. Mutation of beta-glucosidase 2 causes glycolipid storage disease and impaired male fertility., *J. Clin. Invest.*, 116 (2006), 2985–94.

[117] Santinha, D; Ferreira-Fernandes, E; Melo, T; Silva, EMP; Maciel, E; Fardilha, M; Domingues, P; Domingues, MRM. Evaluation of the photooxidation of galactosyl- and lactosylceramide by electrospray ionization mass spectrometry., Rapid Commun. *Mass Spectrom.*, 28 (2014), 2275–84.

[118] Levery, SB; Glycosphingolipid structural analysis and glycosphingolipidomics., *Methods Enzymol.*, 405 (2005) 300–69.

Chapter 7

AN OVERVIEW OF SPHINGOLIPIDS AND CERAMIDES OF AQUEOUS HUMOR AND TRABECULAR MESHWORK

Yenifer Guerra[1], Katyayini Aribindi[1], Genea Edwards[1,2], Hernando J. Olivos[3], Tasneem M. Bahrainwala[3], Nawajes A. Mandal[4] and Sanjoy K. Bhattacharya[1,2,*]

[1]Bascom Palmer Eye Institute, University of Miami, Miami, FL, US
[2]Department of Biochemistry and Molecular Biology, University of Miami, Miami, FL, US
[3]Waters Corporation, Beverly, MA, US
[4]Dean McGee Eye Institute, University of Oklahoma, Norman, OK, US

ABSTRACT

We present a review of recently analyzed sphingolipid and ceramide profiles of human control and primary open angle glaucoma (POAG) trabecular meshwork (TM) and aqueous humor (AH). The following three classes of lipids: sphingomyelin, sphingoid base, sphingoid base-1-phosphate have been collectively termed as sphingolipids here. The original published profiles were generated by extensive class specific mass spectrometry. Comparisons of respective profiles between control and POAG (TM or AH) showed a majority of sphingolipids and ceramides to be common and very few unique species associated with a given state (control or diseased state) between these two groups. In human TM, sphingomyelin (SM) species were higher in control than in POAG, while the opposite was observed in AH. In human controls, sphingoid base (SB) species and ceramides (Cer) were higher in both TM and AH. Sphingoid base-1-phosphates (SB1P) was found elevated in control AH but the levels remained the same for control and POAG in the TM.

* Corresponding author: Sanjoy K. Bhattacharya. Bascom Palmer Eye Institute, University of Miami, Miami, FL 33136 US. Department of Biochemistry and Molecular Biology, University of Miami, Miami, FL 33136, US. Telephone: 305-482-4103, E-mail: sbhattacharya@med.miami.edu.

In order to test for changes in the sphingolipid and ceramide profiles between normotensive and hypertensive states we repeated the analyses with DBA/2J mice (a mouse model of glaucoma, which undergoes spontaneous transition from normotensive to hypertensive state). We compared TM and AH from the normotensive (mice with an IOP of $\leq$ 17 mmHg) and the hypertensive state (with an IOP of > 18 mmHg). Since the normotensive mice are younger in age (3-6 months) than those in the hypertensive state (8-10 months in age) thus the aging changes are not completely ruled out from these studies. The sphingolipid and ceramide profiles of hypertensive state DBA/2J showed similarity with human POAG with respect to both the TM and AH. The male and female DBA/2J mice showed variation in their lipid composition of TM and AH, but the overall trend of each of the four lipid species showed similar changes. SM and SB species were higher in the normotensive state for both AH and TM. SB1P and ceramides were lower in normotensive .AH and TM. The comparison of human control, POAG and mice normotensive, hypertensive states gives insight into the changes that occur in the hypertensive state without any drug treatment for lowering intraocular pressure. We also present preliminary imaging mass spectrometry (IMS) data for sphingomyelin. Our current IMS results are consistent with extractive mass spectrometry findings.

INTRODUCTION

Glaucoma is a leading cause of irreversible blindness worldwide characterized by optic nerve degeneration [1]. Primary open angle glaucoma (POAG) is the most prevalent form of glaucoma and is most frequently due to increased intraocular pressure (IOP). In the normal eye aqueous humor (AH), a clear fluid found in the anterior eye chamber (before lens) helps maintain the metabolic functions of avascular structures of the anterior eye such as cornea and the trabecular meshwork (TM) and also helps maintain the IOP.

The AH normally exits the anterior chamber of the eye usually through the TM into the Schlemm's canal (SC), which is termed as conventional AH outflow route. Some AH also exits through the ciliary body and other tissues and this route of AH outflow is termed as uveoscleral [2, 3]. In POAG, the IOP is elevated due to impeded outflow consistent with an increased resistance to AH outflow at the level of TM. The loss of vision in glaucoma is attributed to an increase in IOP and subsequent damage to the optic nerve. The abnormal increased resistance to AH outflow remains poorly understood. It is conceivable that the outer boundaries of cells that populate the TM (and also possibly other regions of anterior chamber) must undergo changes during the initial IOP dysfunction and subsequent period of altered IOP homeostasis. The outer boundaries of all cell membranes harbor lipids.

However, aside from serving as constituents of membrane cytoskeleton, lipids are also important signaling molecules. Lipids have been found to play an important role in AH regulation and outflow [4]. The lipids of a single subclass eicosanoids (prostanoids) was historically identified in 1955 in the anterior eye chamber (in the iris and hence was named irin) as a potent IOP modulator [5].

Analogues of prostaglandin $F_{2\alpha}$, are currently used to lower IOP by directing aqueous humor drainage through the uveoscleral outflow pathway [6]. Sphingoid base-1-phosphate is another prominent modulator of AH outflow [7, 8]. Until recently the lipids of classes other than eicosanoid subclass were not investigated due to the lack of availability of suitable tools. Knowledge of the changes in lipid composition before and after IOP elevation is essential for better understanding the pathophysiology of glaucoma.

Sphingolipids and ceramides are classes of bioactive lipids that are evolutionarily conserved among all eukaryotes. Although previously considered only structural components of biological membranes, today sphingolipids are known to play important roles in cell function. Sphingomyelins, sphingoid base, sphingoid base-1-phosphate and ceramides have been implicated as modulators of cell growth, apoptosis, cell adhesion, and inflammation [9]. Recent advances in mass spectrometry has allowed for the high throughput identification and quantification [10] of lipids in small samples such as AH and TM.

Human AH and TM are materials to gain insight into the changes that occur in the pathologic state. However, human samples are often limited with respect to untreated samples in the glaucomatous state. Several animal models of glaucoma exist. However, there are very few animal models of glaucoma where IOP elevation is spontaneous. DBA/2J is a well characterized mouse model of glaucoma [11, 12], where IOP elevation is spontaneous, occurs around 8-9 months of age. DBA/2J is a small animal thus easy to maintain, breed and also enables comparison with genetically matched mouse strain such as DBA/2-Gpnmb+-Sj/J [13]. The later does not develop high IOP or glaucomatous neurogeneration. In this chapter we present an account of analyses of sphingolipids from human control and POAG subjects as well as that from normotensive and hypertensive states of DBA/2J mice.

METHODS

Original human [14, 15] and DBA/2J [16, 17] sphingolipid and ceramide profiles were generated using a triple quadrupole moderate resolution (1 amu resolution at 0.7 FWHM) mass spectrometer (TSQ Quantum Access Max) in direct infusion mode with a Triversa Nanomate. Ratiometric two step quantification was achieved using appropriate lipid standards. The original published data and supplementary data were used for analyses presented in the current review [14-17].

Imaging Mass Spectrometry

The normotensive (usually 3 month old; intraocular pressure $\leq$ 17 mm of Hg) and hypertensive (usually 8 month old; intraocular pressure > 18 mm of Hg) DBA/2J unfixed eyes were flash frozen in optimum cutting temperature medium and subjected to imaging mass spectrometry (IMS) on a MALDI G2S-HDMS SYNAPT (Waters Corporation, Beverly, MA) for a sphingomyelin species.

The MS/MS neutral loss scan was performed for 213.2 in positive ion mode using a collision energy of 50V. The lipid N-oleoyl-D-erythro-sphingosylphosphorylcholine (catalog number 860587; molecular mass 729.08 from Avanti Polar Lipids, Alabaster, Alabama) was used as a standard for the sphingomyelin class. This lipid was spotted in an adjoining region of the MALDI plate to detect and perform MS/MS in an identical fashion.

RESULTS AND DISCUSSION

Sphingolipids and Ceramides of Human AH

The profiling of sphingolipids (sphingoid base, sphingoid base-1-phosphate, sphingomyelin) and ceramides of human AH has revealed that they undergo a change in the POAG compared to controls [15]. A detailed analyses of different species of these classes suggest that most lipids undergo a change (mostly a reduction) in POAG and very few lipids species appear unique either in control or in POAG [15].

Figure 1 depicts the changes in sphingolipid classes in human control or glaucoma as a percent of total sphingolipid plus ceramides. The control has higher amounts in all these classes of lipids except sphingomyelin, whose percent is higher in the glaucomatous AH. POAG is often associated with stiffness or elevated elastic modulus [18]. Sphingomyelins often are associated with membrane fluidity [19-21]. Membrane fluidity that contributes to the elastic modulus [22, 23] may be affected in glaucoma.

There is a possibility that increased sphingomyelin content in the AH is associated with their respective loss from cell membranes of different tissues of the anterior eye chamber, which ultimately find their way in the AH. Thus compositionally the relative quantities of sphingolipids and ceramides undergo a change but there is little change in lipid entities (very few unique lipids in control or glaucoma). This suggests that the disease is possibly a spectrum of dysfunction of IOP homeostasis (or cellular homeostasis) rather than an absolute or unique state from the lipid composition point of view.

Sphingolipids and Ceramides of Human TM

TM lipids were also analyzed using class specific mass spectrometric techniques [14]. Total lipid changes are depicted in Figure 1 as a heat map. The sphingolipid changes in the TM are the exact opposite for sphingomyelin in the TM than in the AH, which is consistent with loss from the TM making its way into the AH for sphingomyelin. However, for sphingoid base, sphingoid base-1-phosphate and ceramides there is an increased amount in control than in the glaucoma (Figure 1).

The original data is protein normalized therefore it is not entirely explicable on the basis of cell loss from the TM. Though the cell loss from TM will contribute towards a decrease in total amount of specific lipid classes as percent of the total (Figure 1).

If AH and TM existed in a perfect equilibrium then changes in TM lipids would be complemented with that of AH. If there is disruption in TM cells then those lipids are likely to result in the AH thus reduced levels of one or more classes of lipids in the TM should result in increased levels in the AH.

Interestingly the experimental results (Figure 1) do depict a parallel trend with respect to total levels of sphingolipids and ceramides suggesting there is more complexity to this picture. The unique lipids in the TM or AH in the pathologic states are very few as narrated above. The unique lipids of TM and AH are not common (Table 1).

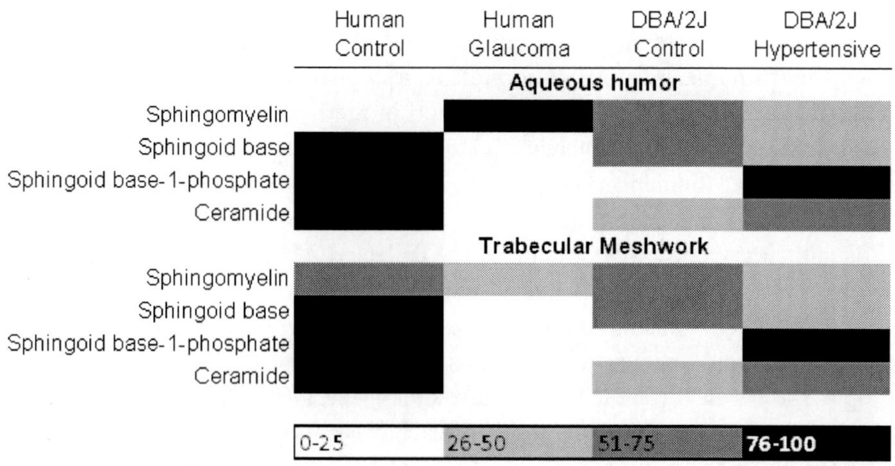

Figure 1. Heat map of total sphingolipids (Sphingomyelin, sphingoid base, sphingoid base-1-phosphate) and ceramides in aqueous humor and trabecular meshwork. The individual class of lipid amount has been expressed as total percentage of total sphingolipids and ceramides for that group of samples (human control or glaucoma or DBA/2J normo- or hypertensive). Scale represents range of percentage as indicated in the bottom using color codes.

Table 1. Unique lipids in Aqueous humor and Trabecular meshwork*

	Aqueous humor				Trabecular meshwork			
Lipid Species	Human Control	Human Glaucoma	DBA/2J Control	DBA/2J Hypertensive	Human Control	Human Glaucoma	DBA/2J Control	DBA/2J Hypertensive
SM(d16:1/17:0)		X						X
Cer(d18:0/24:1(15Z))		X		X				X
SM(d18:1/20:0)				X				X
SM(d18:0/20:0)				X				X
(4E,8Z,d18:2) sphingosine		X						X
Cer(d18:1/26:1(17Z))	X						X	
SM(d18:0/16:0)	X						X	
SM(d18:0/24:0)				X				X
SM(d16:1/18:0)				X				X
SM(d18:1/15:0)				X				X

*Based on preliminary profiling findings using moderate resolution mass spectrometry (references [14-17]).

Sphingolipids and Ceramides of DBA/2J Mouse AH

The DBA/2J sphingolipids and ceramide total amounts were higher in the normotensive state when compared to the hypertensive state [17] except for sphingoid base-1-phosphate and ceramides in the mouse AH (Figure 1).

Expressed as a percent of total lipids (throughout this text, total lipid means total sphingolipids and ceramides and not all lipids of all classes), the trend is a reduction in the control, whether in humans or in the normotensive (control) DBA/2J mice. Sphingomyelin and sphingoid base showed the greatest reduction in the hypertensive state while sphingoid base-1-phosphate and ceramides increased. However, the extent of decrease in hypertensive state is not similar to humans. There are several potential reasons for this: (1) that AH differences are due to changes in the anterior chamber tissues distal to the TM such as lens, cornea, iris and ciliary body; (2) in humans the collected AH is not from hypertensive state but from glaucoma medication (or surgery) induced post IOP elevation state, (3) all analyses are based on moderate resolution extractive mass spectrometry. Possible detection of localization of all lipid species will eventually help understand what are the changes in specific lipid species in different tissues surrounding the AH and also help enable at least semi-quantitative changes in the total lipids of each class and also the total lipids in regional tissues. Advent of imaging mass spectrometry (IMS) is poised to help such analyses in the near future. A preliminary analysis for sphingomyelin showed a decrease in the hypertensive state (Figure 2A-F).

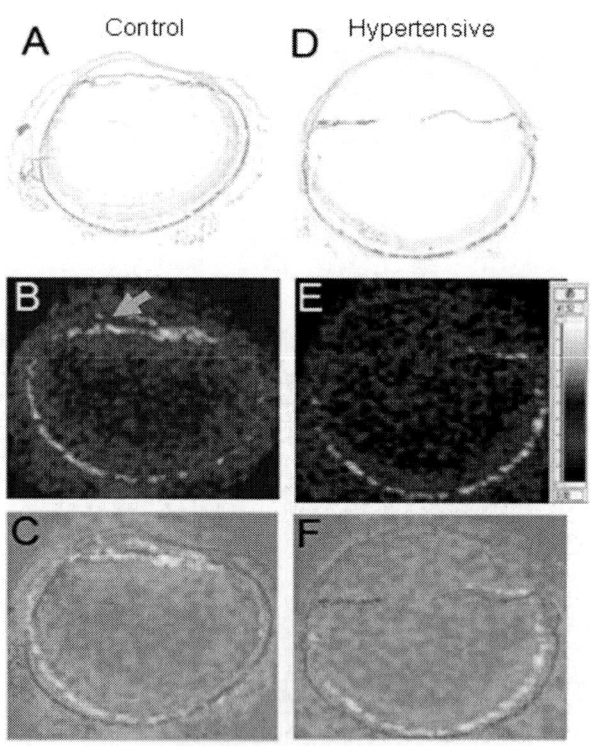

Figure 2. The imaging mass spectrometry (IMS) of a sphingomyelin [SM(d18:1/20:1)]Na+]. The unfixed and snap frozen DBA/2J mouse eyes from A. –C. Control (normotensive) and D. –E. hypertensive state were subjected to careful sectioning (10μm) and IMS. The panels A.. and D. show the Haematoxylin-Eosin stained microscopic images of consecutive sections (5X magnification). The panels B. and E. show the IMS on MALDI G2S-HDMS SYNAPT for a sphingomyelin species SM(d18:1/20:1) or N-(11Z-eicosenoyl)-sphing-4-enine-1-phosphocholine bound with sodium that is [M+Na]+ as m/z of 779.60. This species is dramatically reduced in the hypertensive state corroborating measurements by extractive mass spectrometry. Scale from low (blue) to high (yellow) as indicated. The panels C. and F. are merged microscopic and IMS images.

The hypertensive state showed reduced sphingomyelin species SM(d18:1/20:1). The systematic name for this species is N-(11Z-eicosenoyl)-sphing-4-enine-1-phosphocholine. It was identified in IMS as bound with sodium that is [M + Na]+ therefore showing as a m/z of 779.60. We performed MS/MS analyses for neutral loss scan of 213.2 in positive ion mode to ensure that this indeed is a sphingomyelin species.

This result is consistent with decreased sphingomyelin. The SM(d18:1/20:1) was found with much higher intensity in the residual AH in the normotensive mouse trapped between lens and iris (Figure 2B; indicated by yellow arrow). At present most commercial IMS devices are limited by the resolution of laser, which offers at best about 10 μm spatial resolution. However, for a gross estimation of localization of lipid species in the anterior chamber, IMS with its current limitations is still a good tool and is poised to complement extractive mass spectrometry. Our results from extractive mass spectrometry [17] (Figure 1) are consistent for sphingomyelin with IMS (Figure 2).

Sphingolipids and Ceramides of DBA/2J TM

The sphingolipid profiles of DBA/2J mouse TM [16] revealed a decrease in both sphingomyelin and sphingoid base (Figure 1). However, there was a modest increase in ceramide and a significant increase in sphingoid base-1-phospate (Figure 1). Both females and males showed similar trends for sphingolipids and ceramides [16] in the hypertensive compared to the normotensive state.

A comprehensive analysis of the lipid composition of the AH and TM in both control and glaucoma patients [14, 15] has revealed changes in total sphingolipids and ceramides (Figure 1) as well as some absolute changes (within the sensitivity limits of detection) in a few of these lipid species (Table 1). Since hypertensive state AH or TM cannot be analyzed in humans for ethical reasons, further analysis of the molecular changes under high IOP were conducted on the model organism DBA/2J mice. The DBA/2J mouse presents spontaneous IOP elevation (hypertensive state) around the 8th month that typically prolongs approximately until the 12th month [11, 12]. The lipid profiles for normotensive state (3 months and up to 7.5 months with IOP $\leq$ 17 mm of Hg) were compared to the hypertensive state (~ 8 months IOP > 18 mm of Hg).

The AH is formed by the ciliary processes and its composition changes as it passes the vitreous, lens, iris, and cornea [2]. AH outflow occurs through the uveoscleral and trabecular pathways [2]. In the conventional pathway, AH outflow initiates with the TM then into Schlemm's canal and to the episcleral veins. The uveoscleral outflow pathway is IOP independent unlike the conventional pathway that decreases the outflow as pressure increases. Imbalance between production and outflow usually results in accumulation of AH and results in IOP elevation. Elevated IOP often leads to the onset and progression of glaucoma and it is thus far the only modifiable factor. The impeded outflow has been implicated in the IOP elevation and is attributed to increased resistance to outflow at the level of TM. However, the AH accumulation may depend on several other factors including a compositional change in AH itself.

Current most prominent IOP lowering medications are physiologically active prostaglandin lipid analogues that act by enhancing the unconventional (uveoscleral) outflow [24]. Sphingolipids and ceramides have been known to be involved in several cellular processes and also influence inflammation [25, 26]. They have been implicated in a number of neurodegenerative disorders [25, 27]. They are highly inter-convertible, which happens by the action of several enzymes (Figure 3). Their generation and inter-conversions are highly regulated. Manipulation of these enzymes can result in drastic changes at the tissue or the biological fluids (such as AH) composition of these lipids.

Sphingosine-1-phosphate (SB1P) has shown to decrease IOP in both human [28] and C57BL/6 mice [29]. SB1P decreased the outflow by 39% in mice and 36% in human, suggesting similar pharmacologic signaling mechanism. Studies with the SB1P analogue fingolimod (FTY720) in rats displayed substantial neural preservation in the ganglion cell layer and optic nerve [30]. Induction of nitric oxide synthase (NOS2) has been correlated with neural retinal cell death in elevated IOP. Regulation of the NOS2 pathway through the manipulation of neutral sphingomyelinase (N-SMase) significantly reduced the percent of apoptotic cells in elevated IOP rat model [31].

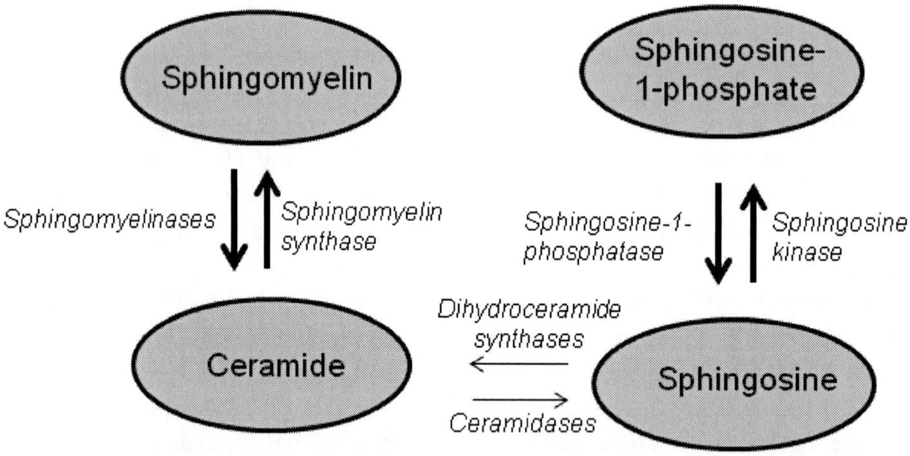

Figure 3. A schematic metabolic pathway diagram. Diagram depicting inter-conversions of different sphingolipid species and the corresponding enzyme (*Italicized*) that catalyzes the reactions.

The lipid profile of human AH revealed several ceramides and sphingolipids unique to either the control or glaucoma states. In control AH, six species were found unique when compared to glaucoma samples. Two lipids, Cer(d18:1/26:1(17Z)) and SM(d18:0/16:0) were unique to human control AH and also in the control DBA2/J TM (Table 1). These species were not found in human glaucoma or mouse hypertensive samples. Their presence in only control subjects suggests that a physiochemical change occurs possibly during the transition in the diseased state. The biochemical changes in AH or TM that is concomitant with the transition towards an ocular hypertensive state remains unknown but a trend seems to emerge from the human and DBA/2J control lipid profiles (Figure 1). Two lipids SM(d16:1/17:0) and (4E,8Z,d18:2) sphingosine were only present in the human glaucoma AH subjects and also in the DBA/2J hypertensive TM (Table 1). Furthermore, Cer(d18:0/24:1(15Z)) was common in human glaucoma AH, DBA2/J hypertensive TM and AH (Table 1). The role played by these species is yet not known but their presence in both humans and the DBA/2J mice reveal a pattern that is likely to hold even with a further expansion of analyses with larger samples.

CONCLUSION

Comparison of the lipid profiles of the human AH and TM in patients without and with glaucoma has revealed species that are common and unique (Table 1) between both groups [14, 15]. Further comparison with the DBA/2J mice model [16, 17] has expanded our understanding of the molecular changes that the AH and TM undergo from the normotensive to the hypertensive state.

The analyses have revealed similar trends between humans and mouse (Figure 1). Preliminary IMS data on specific lipid species (Figure 2) is consistent with the conclusions derived using extractive quantitative mass spectrometry. DBA/2J is a suitable glaucoma model where elevation in IOP is spontaneous and thus provides access to hypertensive state fluid and tissues that cannot be readily obtained from the human subjects. The identified unique sphingolipid and ceramide species might serve as possible markers for glaucoma if they are validated with an expanded number of clinically characterized samples and high resolution mass spectrometry, combined with specific confirmatory chemical modifications.

ACKNOWLEDGMENTS

This work has been partially funded by US Department of Defense grant W81XWH-15-1-0079. Genea Edwards and Yenifer Guerra are supported by a NIH T32 grant EY023194 and a grant from Wallace H. Coulter Center Grant respectively.

REFERENCES

[1] Quigley, H. A., Tielsch, J. M., Katz, J. and Sommer, A. Rate of progression in open-angle glaucoma estimated from cross-sectional prevalence of visual field damage. *Am. J. Ophthalmol.* 122, 355-63 (1996).

[2] Morrison, J. C. and Acott, T. S. In: *Glaucoma Science and Practice* (eds. Morrison, J. C. and Pollack, I. P.) 34-41. (Thieme Medical Publishers Inc., New York, 2003).

[3] Fautsch, M. P. and Johnson, D. H. Aqueous humor outflow: what do we know? Where will it lead us? *Invest. Ophthalmol. Vis. Sci.* 47, 4181-7 (2006).

[4] Wan, Z., Woodward, D. F. and Stamer, W. D. Endogenous Bioactive Lipids and the Regulation of Conventional Outflow Facility. *Expert Rev. Ophthalmol.* 3, 457-470 (2008).

[5] Ambache, N. Irin, a smooth-muscle contracting substance present in rabbit iris. *J. Physiol.* 129, 65-6P (1955).

[6] Toris, C. B., Gabelt, B. T. and Kaufman, P. L. Update on the mechanism of action of topical prostaglandins for intraocular pressure reduction. *Surv. Ophthalmol.* 53 Suppl. 1, S107-20 (2008).

[7] Sumida, G. M. and Stamer, W. D. S1P(2) receptor regulation of sphingosine-1-phosphate effects on conventional outflow physiology. *Am. J. Physiol. Cell Physiol.* 300, C1164-71 (2011).

[8] Sumida, G. M. and Stamer, W. D. Sphingosine-1-phosphate enhancement of cortical actomyosin organization in cultured human Schlemm's canal endothelial cell monolayers. *Invest. Ophthalmol. Vis. Sci.* 51, 6633-8 (2010).

[9] Hannun, Y. A. and Obeid, L. M. Principles of bioactive lipid signalling: lessons from sphingolipids. *Nat. Rev. Mol. Cell. Biol.* 9, 139-50 (2008).

[10] Bhattacharya, S. K. Recent advances in shotgun lipidomics and their implication for vision research and ophthalmology. *Curr. Eye Res.* 38, 417-27 (2013).

[11] John, S. W., Hagaman, J. R., MacTaggart, T. E., Peng, L. and Smithes, O. Intraocular pressure in inbred mouse strains. *Invest. Ophthalmol. Vis. Sci.* 38, 249-53 (1997).

[12] John, S. W., et al. Essential iris atrophy, pigment dispersion, and glaucoma in DBA/2J mice. *Invest. Ophthalmol. Vis. Sci.* 39, 951-62 (1998).

[13] Goel, M., et al. Cochlin, intraocular pressure regulation and mechanosensing. *PLoS One* 7, e34309 (2012).

[14] Aljohani, A., J. et al. Human trabecular meshwork sphingolipid and ceramide profiles and potential latent fungal commensalism. *Invest. Ophthalmol. Vis. Sci.* 55, 3413-22 (2014).

[15] Aljohani, A. J., Munguba, G. C., Guerra, Y., Lee, R. K. and Bhattacharya, S. K. Sphingolipids and ceramides in human aqueous humor. *Mol. Vis.* 19, 1966-84 (2013).

[16] Guerra, Y., Aljohani, A. J., Edwards, G. and Bhattacharya, S. K. A comparison of trabecular meshwork sphingolipids and ceramides of ocular normotensive and hypertensive states of DBA/2J mice. *J. Ocul. Pharmacol. Ther.* 30, 283-90 (2014).

[17] Edwards, G., Aribindi, K., Guerra, Y. and Bhattacharya, S. K. Sphingolipids and ceramides of mouse aqueous humor: Comparative profiles from normotensive and hypertensive DBA/2J mice. *Biochimie* 105, 99-109 (2014).

[18] Russell, P. and Johnson, M. Elastic modulus determination of normal and glaucomatous human trabecular meshwork. *Invest. Ophthalmol. Vis. Sci.* 53, 117 (2012).

[19] Barenholz, Y. and Thompson, T. E. Sphingomyelins in bilayers and biological membranes. *Biochim. Biophys. Acta* 604, 129-58 (1980).

[20] Pedrera, L., et al. Sticholysin I-membrane interaction: an interplay between the presence of sphingomyelin and membrane fluidity. *Biochim. Biophys. Acta* 1838, 1752-9 (2014).

[21] Bruning, B. and Farago, B. Perfluorooctanoic acid rigidifies a model lipid membrane. *Phys. Rev. E Stat. Nonlin. Soft Matter Phys.* 89, 040702 (2014).

[22] Mohandas, N., Chasis, J. A. and Shohet, S. B. The influence of membrane skeleton on red cell deformability, membrane material properties, and shape. *Semin. Hematol.* 20, 225-42 (1983).

[23] Lin, H. K., Zandi, R., Mohideen, U. and Pryadko, L. P. Fluctuation-induced forces between inclusions in a fluid membrane under tension. *Phys. Rev. Lett.* 107, 228104 (2011).

[24] Schachtschabel, U., Lindsey, J. D. and Weinreb, R. N. The mechanism of action of prostaglandins on uveoscleral outflow. *Curr. Opin. Ophthalmol.* 11, 112-5 (2000).

[25] Park, W. J. and Park, J. W. The effect of altered sphingolipid acyl chain length on various disease models. *Biol. Chem.* 396, 693-705 (2015).

[26] Carreira, A. C., Ventura, A. E., Varela, A. R. and Silva, L. C. Tackling the biophysical properties of sphingolipids to decipher their biological roles. *Biol. Chem.* 396, 597-609 (2015).

[27] Ariga, T. Pathogenic role of ganglioside metabolism in neurodegenerative diseases. *J. Neurosci. Res.* 92, 1227-42 (2014).

[28] Stamer, W. D., Read, A. T., Sumida, G. M. and Ethier, C. R. Sphingosine-1-phosphate effects on the inner wall of Schlemm's canal and outflow facility in perfused human eyes. *Exp. Eye Res.* 89, 980-8 (2009).

[29] Boussommier-Calleja, A., et al. Pharmacologic manipulation of conventional outflow facility in ex vivo mouse eyes. *Invest. Ophthalmol. Vis. Sci.* 53, 5838-45 (2012).

[30] You, Y., et al. FTY720 protects retinal ganglion cells in experimental glaucoma. *Invest. Ophthalmol. Vis. Sci.* 55, 3060-6 (2014).

[31] Aslan, M., et al. Inhibition of neutral sphingomyelinase decreases elevated levels of inducible nitric oxide synthase and apoptotic cell death in ocular hypertensive rats. *Toxicol. Appl. Pharmacol.* 280, 389-98 (2014).

Chapter 8

Roles of Sphingolipids in Skeletal Development and Homeostasis

***Garthiga Manickam*[1,*] *and Jingjing Li*[2]**
[1]Faculty of Dentistry, McGill University, Montréal, QC, Canada
[2]Department of Medicine, McGill University, Montréal, QC, Canada

Abstract

Sphingolipids are an integral part of lipid membranes, primarily the plasma membrane of cells. It was thought that these macromolecules were inert structures that were present in cell membranes to maintain cell structure and to serve as adhesion sites for proteins. However, this view has changed throughout the years as recent studies show that sphingolipids and their metabolites have downstream effects on cell signal-transduction pathways, which regulate cell differentiation, proliferation and apoptosis. Recent data also give importance to sphingolipids as regulatory molecules in skeletal tissues. Sphingolipid metabolism is critical for fundamental cellular events and this chapter will discuss current data showing the various roles of some key sphingolipids and their bioactive metabolites in skeletal development and tissue homeostasis.

Introduction

For a long time, the skeleton was considered as a structural and biomechanical support, as well as a reservoir to maintain the homeostasis of two essential mineral ions, inorganic phosphate (P_i) and calcium ions, which are needed for many critical functions in the body. However, more recently, it has been shown that in addition to the roles mentioned above, the skeleton also functions as an endocrine organ, which secretes hormone(s) to regulate energy metabolism and skeletal homeostasis (Karsenty 2003; Karsenty 2011).

* Corresponding author: Garthiga Manickam. Address: Shriners Hospital for Children Montreal, 1529 Cedar Avenue, room no. 333, Montreal, Quebec H3G 1A6, Canada. E-mail: garthiga.manickam@mail.mcgill.ca, phone: (514) 282-7180.

Taking the various roles of the skeleton into consideration, it can be said that it is one of the most dynamic tissues in the body. Our skeleton is made up of unique extracellular matrix components and structurally it is also one of the most complex tissues of the body. At least three different cell types − the chondrocytes (synthesize the cartilage matrix), osteoblasts (synthesize the bone matrix) and osteoclasts (resorb the mineralized matrix) work in concert to achieve the development, remodeling and overall homeostasis of the skeletal tissues. A large number of molecules including autocrine, paracrine and endocrine factors regulate the differentiation of these cells from two different types of stem cells and directly or indirectly affect their functional properties (Karsenty 2003; Karsenty et al. 2009; Mackie et al. 2011; Martin and Seeman 2008). Recently, sphingolipids and their precursor metabolites and breakdown products have been identified as critical regulators of various skeletal cell types (Alebrahim et al. 2014; Aubin et al. 2005; Guenet 1981; Khavandgar et al. 2011; Stoffel 1999; Stoffel et al. 2007; Stoffel et al. 2005). The current chapter will primarily focus on the established roles of some of these sphingolipids and related metabolites/products *in vivo*. Readers interested about the *in vitro* studies that established the role of these molecules in cell culture systems can refer to a recent review article by Khavandgar et al. (Khavandgar and Murshed 2015).

SKELETAL DEVELOPMENT

Skeletal development is a highly regulated multistep process, which involves three major cell types – osteoblasts and osteoclasts in the bone, and chondrocytes in the cartilage.

The precursor cells that give rise to these cell types are distributed in a spatiotemporal manner. The proliferation, differentiation and apoptosis of these cell types are genetically regulated, which ultimately determines the length, shape and load-bearing capacity of the skeleton (Karsenty 2003; Karsenty et al. 2009; Mackie et al. 2011).

The bones that make up the skeleton can be categorized into two groups, namely, endochondral and intramembranous bones. Both bone types initially form with the condensation of mesenchymal stem cells (MSCs). The characteristic differences arise thereafter. For the formation of intramembranous bones, the condensed MSCs first proliferate and then differentiate directly into osteoblasts.

On the other hand, the formation of endochondral bones require an initial cartilage 'precast' or 'anlagen,' so in that case, condensed MSCs proliferate and differentiate into chondrocytes first. These chondrocytes then go through further differentiation, giving rise to 4 distinct zones of chondrocytes at different differentiation stages.

The following zones are collectively known as the growth plate: (1) resting zone of chondrocyte precursors; (2) proliferating zone with chondrocytes that produce type II collagen among various other extracellular matrix (ECM) proteins; (3) prehypertrophic zone and (4) hypertrophic zone where the terminally differentiated cells synthesize type X collagen before going through programmed cell death.

The matrix secreted by the terminally differentiated hypertrophic chondrocytes become mineralized, resorbed by osteoclasts and are eventually replaced by type I collagen-rich bone matrix secreted by osteoblasts.

At this stage, a concerted resorption and formation activity takes place by osteoclasts and osteoblasts, respectively, which give the trabecular bones their characteristic features (Figure 1) (Karsenty 2003; Karsenty et al. 2009; Mackie et al. 2011).

MINERALIZATION AND REMODELING: TWO UNIQUE FEATURES OF SKELETAL TISSUE

The skeleton is a mineralized hard tissue, which is a supporting framework of the body. The mineralized ECM in bone serves to protect the vital organs, facilitate locomotion, anchor muscles and it is also a mineral reservoir for two important minerals, namely calcium ions and P_i, which is required for many physiological functions (Horton and Degnin 2009; Karsenty 2011).

The process of skeletal mineralization is highly regulated to restrict tissue mineralization to the bones and teeth in vertebrates. In order for mineralization to take place specifically in these tissues, the environment should be void of mineralization inhibitors (i.e., pyrophosphate, alkaline phosphatase, matrix Gla protein, etc.), there should be a proper scaffold on which the hydroxyapatite [HA; $Ca_5(PO_4)_3(OH)$] crystals will accumulate (i.e., collagen fibrils), and the two mineral ions, P_i and calcium, should be present at physiological concentrations to promote HA crystal growth (Millan 2013; Murshed et al. 2005; Cen et al. 2008; Boonrungsiman et al. 2012). The presence of such a promineralization environment can explain the specificity of mineralization in these tissues only. Proper mineralization of the skeletal tissues is key to its functional role and this process is constantly regulated so that the quality of the bone is maintained.

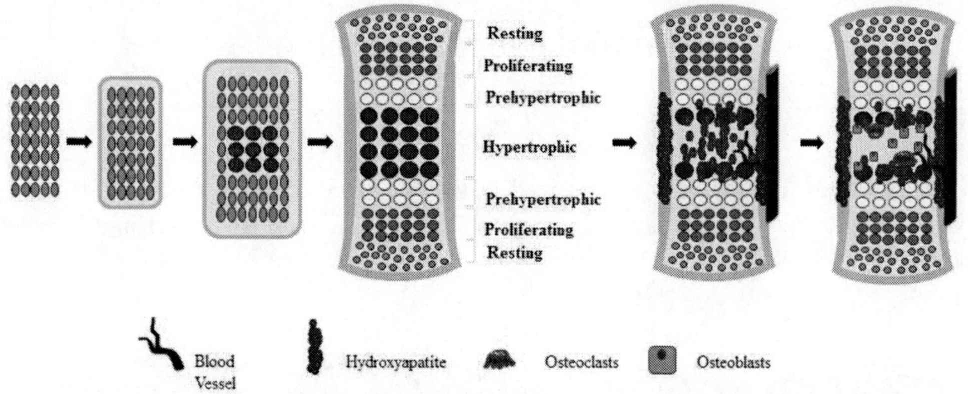

Figure 1. Endochondral ossification begins with the condensation of mesenchymal stem cells, which differentiate to chondrocytes forming the anlagen/precast for the future bone. Chondrocytes at the centre of the precast stop proliferating and become hypertrophic chondrocytes, which secrete a matrix rich in type X collagen. The perichondrial cells differentiate to osteoblasts, forming the bone collar. The matrix secreted by hypertrophic chondrocytes becomes mineralized and blood vessel invasion brings in osteoclast precursors to this region. Mature osteoclasts then resorb the mineralized matrix. Osteoblasts migrate to this region and lay down matrix rich in type I collagen, which gives rise to the primary spongiosa. (Courtesy of Dr. Monzur Murshed, McGill University).

If anything falters in the above-mentioned factors, overall bone quality can be compromised, which may lead to debilitating disorders depending on the severity. Another unique characteristic of the skeleton is that it is being modeled and remodeled almost constantly. These salient features of bone maintain its strength and help repair damages induced by poor load-bearing capacities (DiGirolamo et al. 2012).

The modeling and remodeling process is regulated by a network of cellular machinery comprised primarily of osteoblasts, osteoclasts and osteocytes, the terminally differentiated osteoblasts in the bone (Figure 2) (Seeman 2009). These processes help the bones to adapt to changes in biomechanical forces and to repair damages in the bone, which ultimately results in the maintenance of bone strength (DiGirolamo et al. 2012).

Both the mineralization of skeletal tissues and the modeling/remodeling are regulated by many factors. These regulators include proteins that act as signaling molecules/hormones (e.g., Indian hedgehog, bone morphogenetic proteins, parathyroid hormone, parathyroid hormone-related peptide, etc.) and their respective receptors, transcription factors (e.g., RUNX2, OSX, etc.), enzymes (e.g., PHOSPHO1, TNAP, etc.) and ECM proteins (i.e., Type I collagen, aggrecan, osteocalcin, etc.) (Horton and Degnin 2009; Millan 2013; Karp et al. 2000; Lai and Mitchell 2005; Lombardi et al. 2011; Komori 2003; Sinha and Zhou 2013). With the availability of genetically modified mouse models, the list of factors that regulate skeletal development and tissue homeostasis continues to grow.

More recently, sphingolipids and its related molecules have been shown to take part in various signal transduction pathways to regulate cell proliferation, differentiation, and programmed death and in the mineralization of the skeletal ECM using mouse models (Guenet 1981; Stoffel 1999; Aubin et al. 2005; Khavandgar et al. 2011; Alebrahim et al. 2014).

Sphingolipid Metabolism

Sphingolipids

Sphingolipids are primarily synthesized in the ER of cells. Once synthesized, they are transported into the plasma membrane, where they fulfill their structural and metabolic functions (Gault et al. 2010).

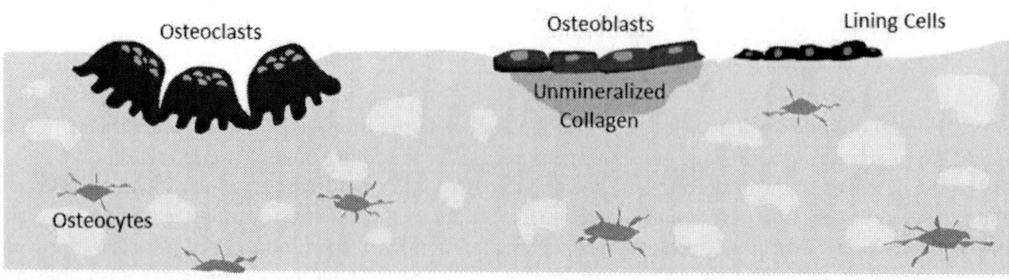

Figure 2. Bone modeling/remodeling showing the different types of cells involved: osteoclasts resorb bone, osteoblasts form bone, and osteocytes (matured osteoblasts trapped in the bone matrix) sense changes in mechanical force on the bone.

Sphingolipids are made up of fatty acids that are linked by amide bonds to a sphingoid base (i.e. sphingosine, sphinganine, or phytosphingosine) (Khavandgar and Murshed 2015). A typical fatty acid chain length ranges from 14-28 carbon atoms and they can be found in saturated or unsaturated forms (Berg 2002). The basic structure of the backbone is an eighteen-carbon amino-alcohol, which is initially synthesized in the endoplasmic reticulum (ER). The modification of this basic backbone structure gives rise to a vast family of sphingolipids, which have diverse structural and functional properties (Figure 3) (Gault et al. 2010). They also give rise to different bioactive metabolites that regulate various cellular functions. Although there are diverse sphingolipids available, each varying in structure and size, their synthesis and catabolism is governed by common pathways.

De Novo Synthesis of Sphingolipids

The *de novo* synthesis of sphingolipids mainly takes place in the ER using simple precursors, such as serine and palmitoyl CoA. Both precursors are used in a condensation reaction catalyzed by an enzyme called *serine palmitoyltransferase* (the rate-limiting enzyme), generating 3-ketosphinganine, which is then converted to sphinganine by *ketosphinganine reductase* (Merrill 2002; Khavandgar and Murshed 2015).

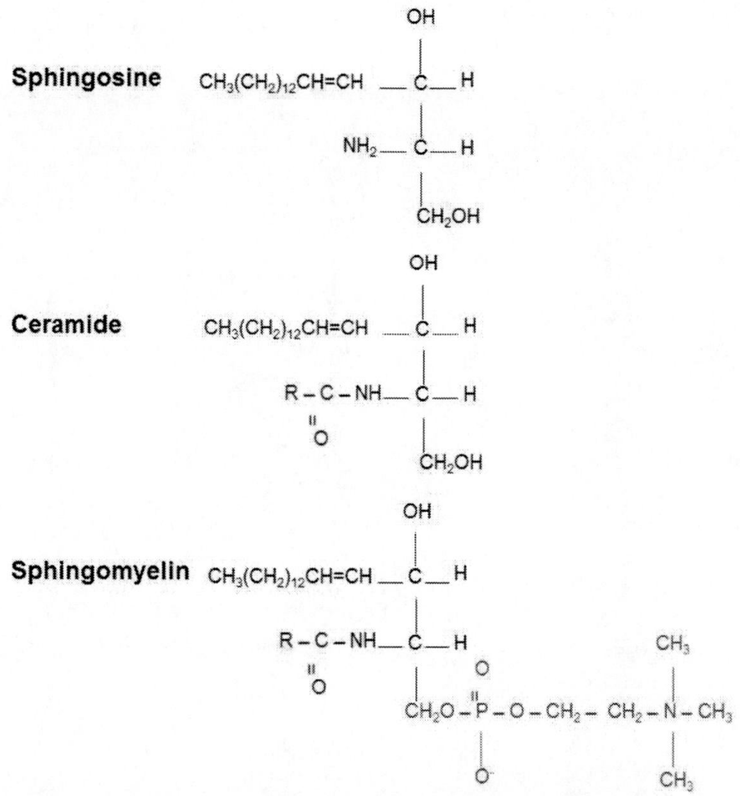

Figure 3. Chemical structures of sphingosine, ceramide and sphingomyelin. Ceramide is composed of sphingosine and a fatty acid (shown in red) which is linked by an amide bond. Sphingomyelin is composed of ceramide and phosphocholine (shown in purple).

Sphinganine is then acetylated by *dihydroxyceramide synthase* to generate dihydroxyceramide, which is then converted by *desaturase* to ceramide, an important second messenger that regulates various cellular functions (Figure 4) (Merrill 2002; Mullen et al. 2012; Khavandgar and Murshed 2015).

Sphingomyelinase Pathway

In addition to the *de novo* synthesis of sphingolipids, there are also various sphingomyelinases (SMase) that play a role in sphingolipid biosynthesis. They cleave sphingomyelin, present in cell membranes, to generate ceramide and phosphocholine (Figure 4). There are three classes of sphingomyelinases – acidic, alkaline and neutral sphingomyelinases. The enzymes are categorized based on the pH in which their enzymatic activity is optimal, or in other words, according to their pH optima (Stoffel 1999; Marchesini and Hannun 2004; Duan 2006).

Among the three classes of sphingomyelinases, acidic sphingomyelinases are the most studied enzymes. There are several studies done on this subclass of enzymes and there are cases where mutations in these enzymes have been shown to be involved in several disorders. For example, mutations in the acidic sphingomyelin phosphodiesterase 1 (*Smpd1*) gene have been shown to be the underlying cause in Niemann-Pick disease (NPD), a lipid storage disease (Rodríguez-Pascau et al. 2009).

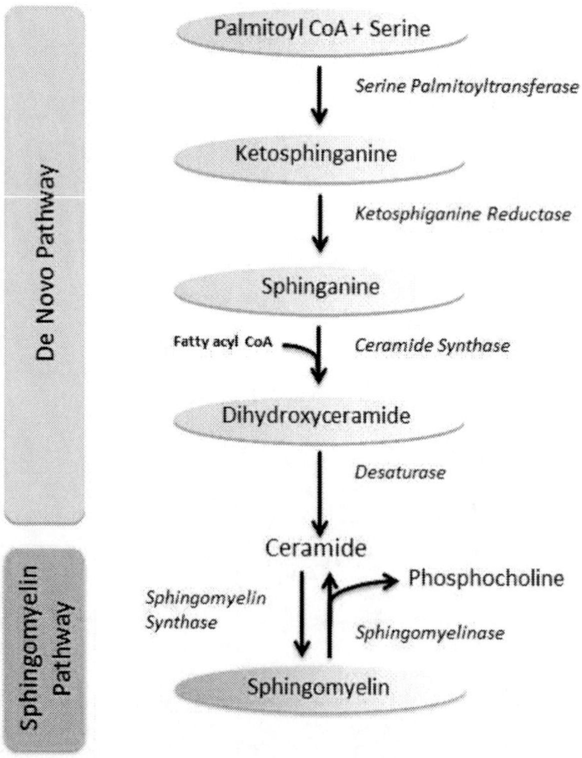

Figure 4. The *de novo* and sphingomyelinase pathways showing ceramide and sphingomyelin metabolism.

This disease is characterized by lipid accumulation in the liver, lung, spleen, bone marrow and brain. There are different forms of NPD associated with *Smpd1* deficiency – NPD type A and NPD type B. Patients with NPD type B have recently been reported to have a skeletal phenotype although the reason for the skeletal manifestation is debatable (i.e., it may be due to the metabolic phenotype itself) (Schuchman and Wasserstein 2015; Wasserstein et al. 2013; Rodríguez-Pascau et al. 2009). However, mouse models for NPD have not been reported to have any skeletal manifestations. There are also recent data suggesting that this enzyme plays many roles, including ceramide-mediated signaling events in various tissues (Gulbins et al. 2013).

Alkaline sphingomyelinases are usually found in the intestinal tract and in human bile. It hydrolyzes sphingomyelin and generates sphingolipid metabolites in the gut. This enzyme has been implicated in colon cancer and intestinal inflammation (Nilsson and Duan 1999). It has been shown that dietary supplementation with sphingomyelin and ceramide analogs inhibit the development of colon cancer in animal models (Nilsson and Duan 1999). It is believed that alkaline sphingomyelinases function as anti-proliferative and anti-inflammatory factors in the gut through the production of ceramide (Duan 2006; Zhang et al. 2008).

Finally, neutral sphingomyelinases are the main enzymes involved in signaling events where the key mediator is ceramide. There are three sphingomyelin phosphodiesterases (SMPD2, SMPD3 and SMPD4) that have been identified as neutral sphingomyelinases (Stoffel 1999). These enzymes are usually membrane-bound, they are highly dependent upon Mg^{2+}, and are found in various tissues – SMPD2 and SMPD4 are expressed in a ubiquitous manner, whereas SMPD3 is mainly expressed in the bone, brain and cartilage (Marchesini et al. 2003).

ROLES OF SPHINGOLIPIDS IN SKELETAL TISSUES

Enzymes involved in the metabolism of two sphingolipids, ceramide and sphingosine 1 phosphate (S1P), have been shown to be critical for the development and homeostasis of the skeletal tissues. These pathways have been studied primarily using *in vivo* mouse models.

Mouse Models with Altered S1P Metabolism and Their Skeletal Phenotypes

S1P, one of the many derivatives of sphingosine (Figure 5), is a well-studied second messenger that has been shown to have a role in skeletal development and tissue homeostasis. Many of the *in vivo* studies done on S1P involve the regulation of the function of osteoclasts and osteoblasts on bone remodeling. A study done by Lotinun et al. showed that osteoclast-derived S1P was increased when the gene encoding cathepsin K, an ECM degrading enzyme secreted by osteoclasts during bone resorption, was inactivated in osteoclasts specifically (Lotinun et al. 2013). This increase in S1P promoted the differentiation of osteoblast precursors to functional osteoblasts, thereby promoting bone formation. The results from this study indicate that osteoclast-derived S1P has the ability to regulate bone remodeling by coupling bone resorption to bone formation. S1P also plays a role in the recruitment of MSCs to migrate and differentiate into functional osteoblasts (Lotinun et al. 2013).

This explains how osteoclasts are able to recruit osteoprogenitors and increase the number of osteoblasts to promote bone formation at sites of bone resorption. In addition to that, S1P is enriched in blood and induces chemotaxis of osteoclast precursor cells to control bone mineral homeostasis (Ishii et al. 2009; Ishii et al. 2010). Furthermore, calcitonin, a hormone, which is primarily known for its pharmacological role as an inhibitor of bone resorption was shown to inhibit the release of S1P from osteoclasts (Keller et al. 2014). This study was primarily done on the observation that although calcitonin was known to inhibit bone resorption, mice that were deficient in calcitonin showed increased bone formation. The study showed that the underlying mechanism for this phenomenon was in fact due to the inhibitory action of calcitonin on the transporter of S1P, inhibiting the release of S1P (Keller et al. 2014). S1P has been shown to (1) induce chemotaxis and recruit both osteoclast and osteoblast precursor cells to sites of bone remodeling, (2) facilitate proliferation, migration and survival of osteoblasts by acting as the coupling factor in the cross-talk between osteoblasts and osteoclasts, and (3) promote bone resorption by promoting osteoclast formation (Fyrst and Saba 2010). This makes it a pharmacologically exploitable factor for bone metabolism disorders. All in all, S1P is now considered as a critical regulator of bone metabolism.

Mouse Models with Altered Sphingomyelin Metabolism and Their Skeletal Phenotypes

The importance of sphingomyelin and its metabolites, ceramide and phosphocholine, in skeletal development primarily arose from studies on two mouse models, which are deficient in neutral sphingomyelinase 2 (nSMase2), also referred to as sphingomyelin phosphodiesterase 3 (SMPD3). Currently, there are only two mouse models that are deficient in SMPD3. One of the mouse models, referred to as *fragilitas ossium (fro)*, carries a chemically induced mutation where a total of 1,758 bps is deleted in the *Smpd3* gene.

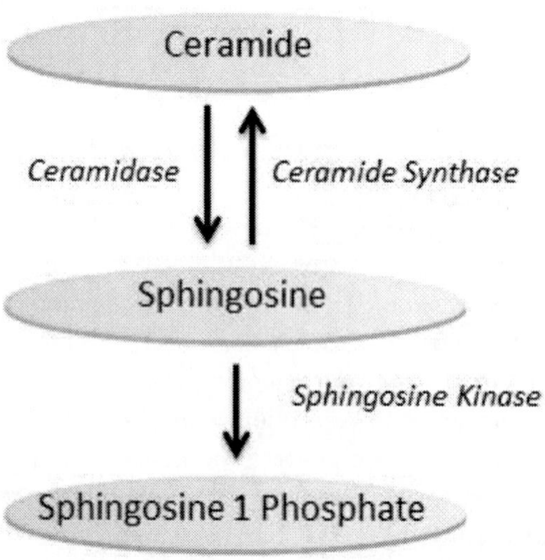

Figure 5. Ceramide is metabolized by *ceramidase* to sphingosine, which is then phosphorylated by *sphingosine kinase* to sphingosine 1 phosphate.

This deletion encompasses part of intron 8 and most of exon 9 of the *Smpd3* gene and renders the enzyme catalytically inactive (Guenet 1981; Aubin et al. 2005). However, the membrane localization of the mutated enzyme is not affected (Khavandgar et al. 2011). The *fro/fro* mice exhibit severe skeletal deformities: they are smaller in size, have bent limbs, and the calvarium is severely undermineralized.

The *fro* mutation was also shown to affect ECM mineralization of the bone in both embryos and adult mice. These mutant mice displayed hypomineralization of the bones, marked by extensive areas of osteoid (unmineralized bone). Additionally, histological sections of the humeri of late-gestational embryos showed that the apoptosis of terminally differentiated hypertrophic chondrocytes was impaired in the growth plate of the *fro/fro* mice and that the cortical bones were poorly mineralized (Khavandgar et al. 2011).

Another mouse model, *Smpd3*$^{-/-}$ knockout mice, was generated using a gene targeting approach in which exon 1 of the *Smpd3* gene was disrupted (Stoffel et al. 2005). SMPD3 deficient *Smpd3*$^{-/-}$ mice were characterized by X-ray imaging and they exhibited a similar skeletal phenotype as the *fro/fro* mice, i.e., short stature and deformities in the long bones and abnormalities in growth plate cartilage (Stoffel et al. 2005).

However, the mineralization defects seen in the *fro/fro* mice were not reported in the *Smpd3*$^{-/-}$ mice, which can be attributed to the differences in the analytical methods and/or the difference in the genetic backgrounds of the mice (*fro/fro* mice were maintained as an inbred colony, and *Smpd3*$^{-/-}$ mice were on a mixed background).

We believe the impaired function of SMPD3 contributes to the skeletal deformities in *fro/fro* mice and that *Smpd3* plays a local and specific role, not a systemic role as suggested by Stoffel et al. in the regulation of skeletal development. They reported a complete rescue of the *Smpd3*$^{-/-}$ mouse from severe skeletal dysplasia by a *Col2a1-Smpd3* transgenic model (Stoffel et al. 2007). However, the transgene is not chondrocyte-specific as claimed and "leaky" expression was seen in almost all tissues (Stoffel et al. 2007).

In a recent study we observed the rescue of the skeletal deformities in *fro/fro;Col1a1-Smpd3* mice where the *Smpd3* transgene is specifically expressed in osteoblasts (Khavandgar et al. 2011). In this transgenic model, the bone defect phenotypes such as the short and bent limbs, and poor bone mineralization were restored. Whereas, the cartilage phenotype that appears early on during skeletal development remained unaffected (Khavandgar et al. 2011).

Interestingly, the most common parameters affecting ECM mineralization such as serum calcium, P_i and alkaline phosphatase levels, were not decreased in *fro/fro* mice. Additionally, the levels of mineralization inhibitors such as matrix-Gla-protein (MGP) or pyrophosphate (PP_i) were not elevated in *fro/fro* mice (Khavandgar et al. 2011). Therefore, *Smpd3* plays a tissue-specific role in regulating ECM mineralization in skeletal tissues through a yet unknown mechanism.

CONCLUSION

Currently, there are not many mouse models that are available to study the role of sphingolipids and its metabolites on skeletal development and tissue homeostasis. There are however a few existing mouse models that were used to elucidate the role of the enzymes from the *de novo* pathway for ceramide synthesis (i.e., *Des1*$^{-/-}$; *desaturase 1* knockout), which can be examined for skeletal phenotypes. This would widen the horizon for the role of bioactive metabolites and/or the enzymes of sphingolipid metabolism on skeletogenesis. Most of the current findings are obtained from *in vitro* analyses using cell lines and primary cells. In order to better understand the role of sphingolipids in skeletal biology, tissue-specific knockouts of the regulatory enzymes with known or possible roles in the skeletal tissues need to be generated and characterized. Furthermore, emerging data also points toward the regulation of energy metabolism through the action of sphingolipids in the skeletal tissue. This is another field of study that can broaden the scope of sphingolipid metabolism in skeletal tissue to other physiological roles.

REFERENCES

Alebrahim, S., Khavandgar, Z., Marulanda, J. and Murshed, M. (2014). Inducible transient expression of Smpd3 prevents early lethality in fro/fro mice. *Genesis*, 52(5), 408-416. doi: 10.1002/dvg.22765.

Aubin, I., Adams, C. P., Opsahl, S., Septier, D., Bishop, C. E., Auge, N., ... Poirier, C. (2005). A deletion in the gene encoding sphingomyelin phosphodiesterase 3 (Smpd3) results in osteogenesis and dentinogenesis imperfecta in the mouse. *Nat. Genet.*, 37(8), 803-805. doi: 10.1038/ng1603.

Berg JM, T. J., Stryer, L. (2002). *Biochemistry Fatty Acids Are Key Constituents of Lipids*. (pp. Section 12.12). Retrieved from http://www.ncbi.nlm.nih.gov/books/NBK22497/.

Boonrungsiman, S., Gentleman, E., Carzaniga, R., Evans, N. D., McComb, D. W., Porter, A. E. and Stevens, M. M. (2012). The role of intracellular calcium phosphate in osteoblast-mediated bone apatite formation. *Proc. Natl. Acad. Sci. US*, 109(35), 14170-14175. doi: 10.1073/pnas.1208916109.

Cen, L., Liu, W., Cui, L., Zhang, W. and Cao, Y. (2008). Collagen Tissue Engineering: Development of Novel Biomaterials and Applications. *Pediatr. Res.*, 63(5), 492-496.

DiGirolamo, D. J., Clemens, T. L. and Kousteni, S. (2012). The skeleton as an endocrine organ. *Nat. Rev. Rheumatol.*, 8(11), 674-683. doi: 10.1038/nrrheum.2012.157.

Duan, R. D. (2006). Alkaline sphingomyelinase: an old enzyme with novel implications. *Biochim. Biophys. Acta*, 1761(3), 281-291. doi: 10.1016/j.bbalip.2006.03.007.

Fyrst, H. and Saba, J. D. (2010). An update on sphingosine-1-phosphate and other sphingolipid mediators. *Nat. Chem. Biol.*, 6(7), 489-497. doi: 10.1038/nchembio.392.

Gault, C. R., Obeid, L. M. and Hannun, Y. A. (2010). An overview of sphingolipid metabolism: from synthesis to breakdown. *Adv. Exp. Med. Biol.*, 688, 1-23.

Guenet, J. L., Stanescu, R., Maroteaux, P., Stanescu, V. (1981). <Fragilitas ossium - a new autosomal recessive mutation in the mouse.pdf>. *J. Hered.*, 72(6), 440-441.

Gulbins, E., Palmada, M., Reichel, M., Luth, A., Bohmer, C., Amato, D., ... Kornhuber, J. (2013). Acid sphingomyelinase-ceramide system mediates effects of antidepressant drugs. *Nat. Med.*, 19(7), 934-938. doi: 10.1038/nm.3214.

Horton, W. A. and Degnin, C. R. (2009). FGFs in endochondral skeletal development. *Trends Endocrinol. Metab.*, 20(7), 341-348. doi: 10.1016/j.tem.2009.04.003.

Ishii, M., Egen, J. G., Klauschen, F., Meier-Schellersheim, M., Saeki, Y., Vacher, J., ... Germain, R. N. (2009). Sphingosine-1-phosphate mobilizes osteoclast precursors and regulates bone homeostasis. *Nature*, 458(7237), 524-528. doi: 10.1038/nature07713.

Ishii, M., Kikuta, J., Shimazu, Y., Meier-Schellersheim, M. and Germain, R. N. (2010). Chemorepulsion by blood S1P regulates osteoclast precursor mobilization and bone remodeling in vivo. *J. Exp. Med.*, 207(13), 2793-2798. doi: 10.1084/jem.20101474.

Karp, S. J., Schipani, E., St-Jacques, B., Hunzelman, J., Kronenberg, H. and McMahon, A. P. (2000). Indian hedgehog coordinates endochondral bone growth and morphogenesis via parathyroid hormone related-protein-dependent and -independent pathways. *Development*, 127(3), 543-548.

Karsenty, G. (2003). The complexities of skeletal biology. *Nature*, 423(6937), 316-318.

Karsenty, G. (2011). Bone endocrine regulation of energy metabolism and male reproduction. *C R Biol.*, 334(10), 720-724. doi: 10.1016/j.crvi.2011.07.007.

Karsenty, G., Kronenberg, H. M. and Settembre, C. (2009). Genetic control of bone formation. *Annu. Rev. Cell Dev. Biol.*, 25, 629-648. doi: 10.1146/annurev.cellbio.042308.113308.

Keller, J., Catala-Lehnen, P., Huebner, A. K., Jeschke, A., Heckt, T., Lueth, A., ... Amling, M. (2014). Calcitonin controls bone formation by inhibiting the release of sphingosine 1-phosphate from osteoclasts. *Nat. Commun.*, 5, 5215. doi: 10.1038/ncomms6215.

Khavandgar, Z. and Murshed, M. (2015). Sphingolipid metabolism and its role in the skeletal tissues. *Cell. Mol. Life Sci.*, 72(5), 959-969. doi: 10.1007/s00018-014-1778-x.

Khavandgar, Z., Poirier, C., Clarke, C. J., Li, J., Wang, N., McKee, M. D., ... Murshed, M. (2011). A cell-autonomous requirement for neutral sphingomyelinase 2 in bone mineralization. *J. Cell Biol.*, 194(2), 277-289. doi: 10.1083/jcb.201102051.

Komori, T. (2003). Requisite roles of Runx2 and Cbfb in skeletal development. *J. Bone Miner. Metab.*, 21(4), 193-197. doi: 10.1007/s00774-002-0408-0.

Lai, L. P. and Mitchell, J. (2005). Indian hedgehog: its roles and regulation in endochondral bone development. *J. Cell. Biochem.*, 96(6), 1163-1173. doi: 10.1002/jcb.20635.

Lombardi, G., Di Somma, C., Rubino, M., Faggiano, A., Vuolo, L., Guerra, E., ... Colao, A. (2011). The roles of parathyroid hormone in bone remodeling: prospects for novel therapeutics. *J. Endocrinol. Invest.*, 34(7 Suppl.), 18-22.

Lotinun, S., Kiviranta, R., Matsubara, T., Alzate, J. A., Neff, L., Luth, A., ... Baron, R. (2013). Osteoclast-specific cathepsin K deletion stimulates S1P-dependent bone formation. *J. Clin. Invest.*, 123(2), 666-681. doi: 10.1172/JCI64840.

Mackie, E. J., Tatarczuch, L. and Mirams, M. (2011). The skeleton: a multi-functional complex organ: the growth plate chondrocyte and endochondral ossification. *J. Endocrinol.*, 211(2), 109-121. doi: 10.1530/JOE-11-0048.

Marchesini, N. and Hannun, Y. A. (2004). Acid and neutral sphingomyelinases: roles and mechanisms of regulation. *Biochem. Cell Biol.*, 82(1), 27-44. doi: 10.1139/o03-091.

Marchesini, N., Luberto, C. and Hannun, Y. A. (2003). Biochemical properties of mammalian neutral sphingomyelinase 2 and its role in sphingolipid metabolism. *J. Biol. Chem.*, 278(16), 13775-13783. doi: 10.1074/jbc.M212262200.

Martin, T. J. and Seeman, E. (2008). Bone remodelling: its local regulation and the emergence of bone fragility. *Best Pract. Res. Clin. Endocrinol. Metab.*, 22(5), 701-722. doi: 10.1016/j.beem.2008.07.006.

Merrill, A. H., Jr. (2002). De novo sphingolipid biosynthesis: a necessary, but dangerous, pathway. *J. Biol. Chem.*, 277(29), 25843-25846. doi: 10.1074/jbc.R200009200.

Millan, J. L. (2013). The role of phosphatases in the initiation of skeletal mineralization. *Calcif. Tissue Int.*, 93(4), 299-306. doi: 10.1007/s00223-012-9672-8.

Mullen, T. D., Hannun, Y. A. and Obeid, L. M. (2012). Ceramide synthases at the centre of sphingolipid metabolism and biology. *Biochem. J.*, 441(3), 789-802. doi: 10.1042/bj 20111626.

Murshed, M., Harmey, D., Millan, J. L., McKee, M. D. and Karsenty, G. (2005). Unique coexpression in osteoblasts of broadly expressed genes accounts for the spatial restriction of ECM mineralization to bone. *Genes Dev.*, 19(9), 1093-1104. doi: 10.1101/gad. 1276205.

Nilsson, A. and Duan, R. D. (1999). Alkaline sphingomyelinases and ceramidases of the gastrointestinal tract. *Chem. Phys. Lipids*, 102(1-2), 97-105.

Rodríguez-Pascau, L., Gort, L., Schuchman, E. H., Vilageliu, L., Grinberg, D. and Chabás, A. (2009). Identification and characterization of SMPD1 mutations causing Niemann-Pick types A and B in Spanish Patients. *Human mutation*, 30(7), 1117-1122. doi: 10.1002/ humu.21018.

Schuchman, E. H. and Wasserstein, M. P. (2015). Types A and B Niemann-Pick disease. *Best Pract. Res. Clin. Endocrinol. Metab.*, 29(2), 237-247. doi: 10.1016/j.beem.2014.10.002.

Seeman, E. (2009). Bone modeling and remodeling. *Crit. Rev. Eukaryot. Gene Expr.*, 19(3), 219-233.

Sinha, K. M. and Zhou, X. (2013). Genetic and molecular control of Osterix in skeletal formation. *J. Cell. Biochem.*, 114(5), 975-984. doi: 10.1002/jcb.24439.

Stoffel, W. (1999). Functional analysis of acid and neutral sphingomyelinases in vitro and in vivo. *Chem. Phys. Lipids*, 102(1-2), 107-121.

Stoffel, W., Jenke, B., Block, B., Zumbansen, M. and Koebke, J. (2005). Neutral sphingomyelinase 2 (smpd3) in the control of postnatal growth and development. *Proc. Natl. Acad. Sci. US*, 102(12), 4554-4559. doi: 10.1073/pnas.0406380102.

Stoffel, W., Jenke, B., Holz, B., Binczek, E., Gunter, R. H., Knifka, J., ... Niehoff, A. (2007). Neutral sphingomyelinase (SMPD3) deficiency causes a novel form of chondrodysplasia and dwarfism that is rescued by Col2A1-driven smpd3 transgene expression. *Am. J. Pathol.*, 171(1), 153-161. doi: 10.2353/ajpath.2007.061285.

Wasserstein, M., Godbold, J. and McGovern, M. M. (2013). Skeletal manifestations in pediatric and adult patients with Niemann Pick disease type B. *J. Inherit. Metab. Dis.*, 36 (1), 123-127. doi: 10.1007/s10545-012-9503-0.

Zhang, P., Li, B., Gao, S. and Duan, R. D. (2008). Dietary sphingomyelin inhibits colonic tumorigenesis with an up-regulation of alkaline sphingomyelinase expression in ICR mice. *Anticancer Res.*, 28(6a), 3631-3635.

Chapter 9

THE EFFECT OF MEMBRANE MICRODOMAINS CONTAINING SPHINGOLIPIDS ON LIPID PEROXIDATION

Angel Catalá*

Instituto de Investigaciones Fisicoquímicas Teóricas y Aplicadas, (INIFTA-CCT La Plata-CONICET), Facultad de Ciencias Exactas, Universidad Nacional de La Plata, Argentina

ABSTRACT

Sphingomyelin (SM) is a major phospholipid in cell membranes and lipoproteins of all animals. Its concentration fluctuates considerably between different cells, and even among membranes of the same cell. An interesting characteristic of SM distribution is that is similar the distribution of free (unesterified) cholesterol (FC). For instance, both these lipids are concentrated in the plasma membrane of the mammalian cells, and are important constituents of the membrane rafts which are involved in a broad diversity of cellular functions. Lipid rafts are microdomains of the plasma membrane that contain high levels of cholesterol and sphingolipids. In terms of lipids, the heterogeneous membrane is believed to consist of a mixture of a dispersed ''lipid raft'' phase, enriched in cholesterol, raft-associated proteins, and saturated lipids (such as sphingolipids), and the ''non-raft'' matrix phase enriched with phospholipids containing polyunsaturated fatty acids (PUFAs) and vitamin E. Phospholipids are essential components of cell membranes that maintain structure and control the function of cells. Lipid peroxidation is one of the major outcomes of free radical-mediated injury to tissue. Peroxidation of fatty acyl groups occurs mostly in membrane phospholipids located in the ''non-raft'' matrix phase. The dispersed ''lipid raft'' phase, enriched in cholesterol, raft-associated proteins, and sphingolipids is not affected by lipid peroxidation because sphingolipids contain only saturated fatty acids in its molecule, Peroxidation of lipids can greatly alter the physicochemical properties of membrane lipid bilayers, resulting in severe cellular dysfunction. In addition, a variety of lipid by products is produced as a consequence of peroxidation. Lipids containing polyunsaturated fatty acids are subject to free radical–

* Corresponding author: catala@inifta.unlp.edu.ar.

initiated oxidation and can contribute in chain reactions that amplify damage to biomolecules.

Keywords: lipid peroxidation, membranes, PUFAs: lipid rafts

Abbreviations

DAPC	1,2-diarachidonylphosphatidylcholine
PDPC	1-palmitoyl-docosahexaenoylphosphatidylcholine
POPC	1-palmitoyl-2-oleoylphosphatidylcholine
AA	arachidonic acid
DHA	docosahexaenoic acid
EPA	eicosapentaenoic acid
FMM	Fluid Mosaic Model
FC	free (unesterified) cholesterol
LO•	lipid alkoxyl radical
LOOH	lipid hydroperoxide
LOO•	lipid peroxyl radical
L•	lipid radical
ld	liquid-disordered phase;
lo	liquid-ordered phase;
NMR	nuclear magnetic resonance
oxPl	oxidized phospholipids
PC	phosphatidylcholine
PE	phosphatidylethanolamine
PUFAs	polyunsaturated fatty acids
SM	Sphingomyelin

1. Fluid Mosaic Model (FMM) of Biological Membrane Structure

When the Fluid Mosaic Model (FMM) of biological membrane structure was introduced 42 years ago, it was visualized as a basic model for cell membranes that could explain existing data on membrane proteins and lipid structures and their dynamics. The mechanisms and principles establishing the thousands of proteins and lipids that make up membrane bilayers in cells are still unclear. Mueller et al. reviewed the basic properties of biological membranes and the most common theories for lateral segregation of membrane components before discussing an emerging model of a self-organized, multi-domain membrane or "patchwork membrane" (Mueller et al. 2012). One of the most interesting questions about cell membranes is why these supra molecular assembly's enclose so many diverse kinds and species of lipids. In any case, if the principal reason of membranes is to compartmentalize cells while serving as selective permeability barriers in the export and import of a variety of cellular substances, then only a bunch of lipids should be required. With the understanding that many membrane lipids themselves serve as reservoirs of metabolically derived messenger signals, the original question appears much less discouraging. Yet, to efficiently control such a complex assembly requires an intrinsic organization much greater than originally imagined when the bilayer lipids were thought to function mainly as a solution to facilitate protein diffusion (Singer and Nicolson 1972). Hence, current advances in the lipid cell signaling field not only have offered clarification regarding the need for variety among membrane lipids but also have served to renovate attention in the physical organization of lipids within biomembranes (Vaz and Almeida 1993; Jacobson et al. 1995). In particular, the postulated incidence of sphingolipids-cholesterol rafts or microdomains in biomembranes has been at the vanguard of new advances in the membrane trafficking and signaling field.

2. The Functions of Many Lipids

The eukaryotic cells use approximately 5% of their genes to produce thousands of diverse lipids. Although we currently differentiate the exact functions of many lipids, the complete utility of the eukaryotic lipid diversity remains unidentified. Lipids achieve three general purposes. First, because of their reduced state, lipids are utilized for energy storage, mostly as triacylglycerol and steryl esters, in lipid droplets.

These function mainly as anhydrous reservoirs for the efficient storage of caloric reserves and as supplies of fatty acid and sterol components that are required for membrane biogenesis. Second, the environment of cellular membranes is formed by polar lipids, which consist of a hydrophobic and a hydrophilic component. The propensity of the hydrophobic moieties to self-associate (entropically established by water), and the tendency of the hydrophilic moieties to interact with aqueous environments and with each other, is the physical origin of the spontaneous arrangement of membranes. This basic principle of amphipathic lipids is a chemical property that allowed the first cells to separate their internal constituents from the external environment. This similar principle is applied within the cell to generate distinct organelles.

This compartmentalization facilitates separation of specific chemical reactions for the purposes of expanded biochemical effectiveness and limited spread of reaction products. In addition to the barrier reason, lipids provide membranes with the potential for budding, fission and fusion, characteristics that are indispensable for cell division, biological reproduction and intracellular membrane trafficking. Lipids also allow particular proteins in membranes to aggregate, and others to disperse. Finally, lipids can act as first and second messengers in signal transduction and molecular recognition processes. The degradation of amphipathic lipids allows for bipartite signaling phenomenon, which can be transmitted within a membrane by hydrophobic portions of the molecule and also spread through the cytoplasm by soluble (polar) portions of the molecule.

Additionally, numerous lipids function to characterize membrane domains, which employ proteins from the cytoplasm that subsequently organize secondary signaling or effectors complexes.

3. Sphingomyelin

Sphingomyelin (SM) is a major phospholipid in cell membranes and lipoproteins of all animals. Its concentration varies considerably among different cells, and even among membranes of the same cell (Barenholz 1984; Slotte 1999). An intriguing feature of SM distribution is that it parallels the distribution of free (unesterified) cholesterol (FC). For example, both these lipids are concentrated in the plasma membrane of the mammalian cells (Lange et al. 1989), and are critical components of the membrane rafts which are implicated in a wide variety of cellular functions (Pike 2003).

4. Lipid Rafts

The lipid molecules of biological plasma membranes are not regularly distributed. They are thought to contain "rafts" – micro domains enriched in sphingolipids and cholesterol. In model membranes, the lipid raft phase is typically associated with a so-called ''liquid-ordered" (lo) phase, while the non-raft phase has been identified as the ''liquid-disordered" (ld) phase based on differences in the short-ranged lipid translational and conformational order (Simons and Vaz 2004). Lipid rafts are dynamic and highly organized membrane nano-domains rich in cholesterol and sphingolipids that are different from adjacent membranes of unsaturated phospholipids. The average size of lipid rafts is estimated to be 50 nm in diameter, although several different raft domains can exist in a cell, which is heterogeneous in size and life time (Simons and Toomre 2000). Lipid rafts concentrate select proteins and serve as a platform for cellular processes such as cell signaling, pathogen entry, cell adhesion, motility, protein sorting and trafficking (Simons and Toomre 2000: Helms and Zurzolo 2004). Although positive identification of rafts in cell membranes has proven elusive, they are usually thought to exist and to concentrate proteins in regulated behavior. When in rafts, proteins could fast interact with each other to perform their functions, and this is thought to occur in processes as varied and far reaching as signal transduction and membrane trafficking (Edidin 2003; He et al. 2005; Parton and Richards 2003, Young et al. 2005). Most studies directed at characterizing rafts in cells have established that they are small, on the order of tens of nanometers (Kusumi and Suzuki 2005; Plowman et al. 2005; Pralle et al. 2000; Sharma et al. 2004). But an ample range of raft sizes has been described (Dietrich et al. 2002; Hess et al. 2005) and rafts may include a vast fraction of the area of a cell membrane (Gidwani et al. 2001).

5. Alpha-Tocopherol and PUFA-Containing Phospholipids Co-Localize in Non-Raft Domains

Vitamin E (alpha-tocopherol) has long been identified as the main antioxidant in biological membranes, and yet several structurally related questions continue of how the vitamin functions. For example, the very low levels of alpha-tocopherol reported for whole cell extracts question how this molecule can effectively defend the relatively enormous quantities of PUFA-containing phospholipids found in membranes that are highly vulnerable to lipid peroxidation attack. Atkinson et al. 2010, have hypothesized that alpha-tocopherol partition into domains that are enriched in polyunsaturated phospholipids, increasing the concentration of the vitamin in the place where it is most required. These highly disordered domains depleted in cholesterol are analogous, but organizationally opposing, to the well-studied lipid rafts. Experimental proof in support of the formation of PUFA-rich domains in model membranes is presented, focusing upon docosahexaenoic acid that is one of the most unsaturated fatty acids commonly found. Physical methodologies are then described to elucidate the nature of the interaction of alpha-tocopherol with PUFA and to establish that the vitamin and PUFA-containing phospholipids co-localize in non-raft domains. Atkinson et al. 2010, have reviewed the status of α-tocopherol's organization in membranes, including vertical location or depth, and have proposed lateral segregation of the vitamin into PUFA-rich domains. These domains are postulated in correspondence with the very well-studied lipid raft domain that at present plays a significant role on membrane structure. While lipid rafts are SM-rich, lo domains, that are held together by cholesterol and have functions relating to cell signaling events, the non-raft domains are PUFA-rich, ld domains in which α-tocopherol co-localizes and serves its antioxidant function. Structural and functional roles for α-tocopherol, thus, would fit together nicely in a reciprocally helpful mode. By tracking after PUFA, the vitamin is situated "in the correct position at the correct time to be an antioxidant." A series of diverse physical methodologies are proposed to test the nature of the hypothesized α-tocopherol/PUFA membrane domains.

6. How Polyunsaturated Fatty Acids Modify Molecular Organization in Membranes

The complexity by which n-3 PUFA can reorganize lipid microdomains emerges from studies on lipid bilayers by solid state 2H NMR spectroscopy supplemented by complementary biophysical techniques. Solid state 2H NMR experiments on PDPE-d31/SM/cholesterol (1:1:1 mol) showed that when DHA is esterified to PE, it segregates away from raft-like regions enriched in SM and cholesterol that are nano-scale in size. This segregation is driven by steric incompatibility between highly disordered DHA chains and the rigid steroid moiety of the sterol that is accentuated with the small head group of PE. 2H NMR spectra for PDPC-d31/SM/cholesterol (1:1:1 mol) indicated, contrastingly, that DHA esterified to PC infiltrates SM-rich/sterol-rich nano-sized domains. These findings provide unprecedented mechanistic insight into results from in vivo experiments on how n-3 PUFA disrupt lipid microdomains. How the introduction of DHA-containing PE into a region of bulk lipid surrounding a raft-like domain has the potential to push cholesterol into the raft and so increase molecular order within the domain is illustrated in a cartoon presented in Figure 9 (upper panel). This scenario, although agreeing with the increased order observed in cells treated with n-3 PUFA, does not reconcile with the redistribution of cholesterol from DRM to DSM seen in cells by detergent extraction. One intriguing possibility is to speculate that n-3 PUFA-containing PE molecules form nanodomains, which is compatible with the size indicated by NMR in model membranes, within a host microdomain that becomes more ordered because sphingolipid as well as cholesterol molecules are expelled. The NMR data establish that n-3 PUFA-containing PC can incorporate into raft-like domains in lipid bilayers, consistent with detergent extraction studies on cells that detected the uptake of EPA and DHA into DRM. What the presence of PDPC or PEPC does to molecular organization within the SM-rich/cholesterol-rich domain formed in the model membrane is yet to be studied and experiments with PSM-d31 are planned to directly probe order therein. It is an issue that has to be addressed since the introduction of n-3 PUFA-containing PC into an ordered raft-like domain, as depicted in a cartoon in Figure 9 (lower panel), would be expected to displace sterol and disorder the domain. The reverse, an increase in order, is typically reported in cells. A possible explanation, interestingly, may lie in the reduction in the concentration of AA in DRM that usually follows elevation in the amount of n-3 PUFA. Order parameters measured for PDPC-d31 are higher (~ 6% in the plateau region) than 1-[2H31] palmitoyl-2-arachidonylphosphatidylcholine ([2H31]16:0–20:4PC, PAPC-d31) [14], so that substituting DHA for AA could be responsible for a rise in order (Shaikh et al. 2015).

7. The Effect of Membrane Microdomains Containing Sphingolipids on Lipid Peroxidation

In terms of lipids, the heterogeneous membrane is believed to consist of a mixture of a dispersed ''lipid raft" phase, enriched in cholesterol, raft-associated proteins, and saturated lipids (such as sphingolipids), and the ''non-raft" matrix phase enriched with phospholipids containing PUFAs and vitamin E. Vitamin E, partition into domains that are enriched in polyunsaturated phospholipids increasing the concentration of the vitamin in the place where take place the lipid peroxidation process and oxidized phospholipids (oxPl) are formed, Figure 1. The phospholipids, containing polyunsaturated fatty acids located in the ''non-raft" matrix are principally susceptible to peroxidation because abstraction from a methylene (–CH2–) group of a hydrogen atom, which holds only one electron, leaves at the back an unpaired electron on the carbon, –•CH–. The occurrence of a double bond in the fatty acid weakens the C–H bonds on the carbon atom near to the double bond and thus facilitates H• subtraction. The initial reaction of OH with polyunsaturated fatty acids produces a lipid radical (L•), which in turn reacts with molecular oxygen to form a lipid peroxyl radical (LOO•). The LOO• can abstract hydrogen from an adjacent fatty acid to produce a lipid hydroperoxide (LOOH) and a second lipid radical (Catala 2006). The LOOH formed can undergo reductive cleavage by reduced metals, such as Fe^{++}, producing lipid alkoxyl radical (LO•). Both alkoxyl and peroxyl radicals stimulate the chain reaction of lipid peroxidation by abstracting other hydrogen atoms (Buettner 1993). Peroxidation of lipids can perturb the assembly of the membrane, causing changes in fluidity and permeability, modifications of ion transport and inhibition of metabolic processes (Nigam and Schewe 2000).

The presence of cholesterol and saturated phospholipids decreases oxygen permeability across the membrane. Alpha-tocopherol, the main antioxidant in biological membranes, partition into domains that are enriched in polyunsaturated phospholipids increasing the concentration of the vitamin in the place where it is most required. On the basis of these observations, it is possible to assume that non-raft domains enriched in phospholipids containing PUFAs and vitamin E will be more accessible by molecular oxygen than lipid raft domains enriched in sphingolipids and cholesterol. This situation will render some nano-domains more sensitive to lipid peroxidation than others.

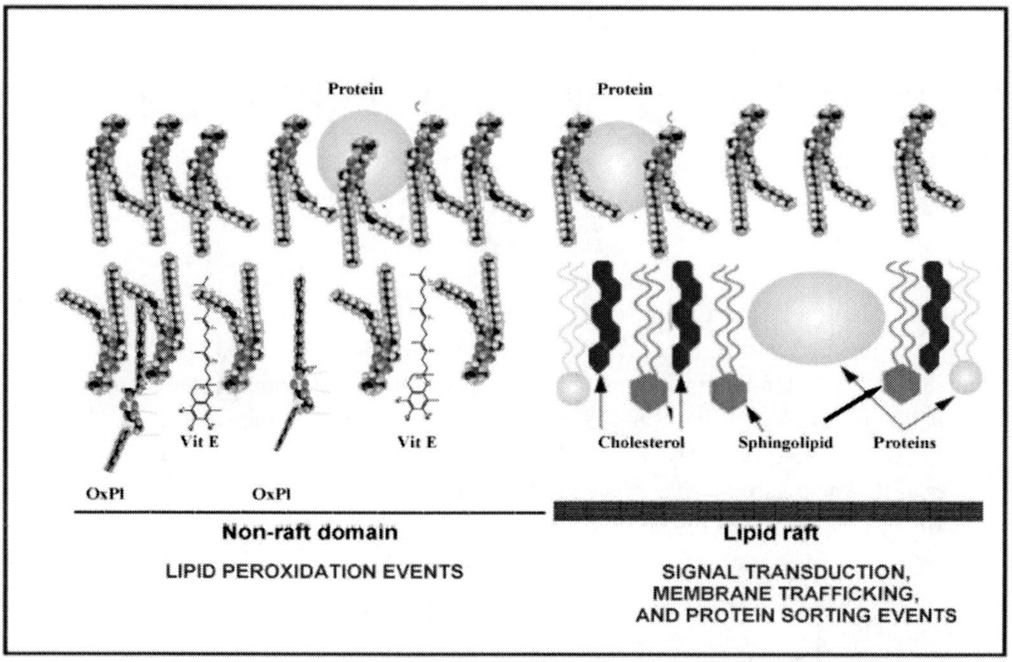

Figure 1. Hypothetical model of the plasma membrane. "Reproduced from A. Catala, Lipid peroxidation modifies the picture of membranes from the "Fluid Mosaic Model" to the "Lipid Whisker Model," Biochimie 94 (2012) 101-109. Copyright © 2012 Elsevier Masson SAS. All rights reserved."

Lipid peroxidation of membrane lipids is accompanied by addition of numerous polar moieties on polyunsaturated fatty acyl chains present in the ''non-raft" matrix phase. Thus, phospholipid oxidation products are very likely to alter the properties of biological membranes, because their polarity and shape may differ significantly from the structures of their parent molecules. Thus, they may modify lipid–lipid and lipid–protein interactions and, as a consequence, also membrane protein functions.

CONCLUSION

The lipid molecules of biological plasma membranes are not regularly distributed. They are thought to contain "rafts" – micro domains enriched in sphingolipids and cholesterol. The presence of cholesterol and saturated phospholipids decreases oxygen permeability across the membrane. Alpha-tocopherol, the main antioxidant in biological membranes, partition into domains that are enriched in polyunsaturated phospholipids increasing the concentration of the vitamin in the place where it is most required. On the basis of these observations, it is possible to assume that non-raft domains enriched in phospholipids containing PUFAs and vitamin E will be more accessible by molecular oxygen than lipid raft domains enriched in sphingolipids and cholesterol. This situation will render some nano-domains more sensitive to lipid peroxidation than others.

Acknowledgments

Studies in the author laboratory were supported by PIP-0157 National Research Council (CONICET).

References

Atkinson, J; Harroun, T; Wassall, SR; Stillwell, W; Katsaras, J. The location and behavior of alpha-tocopherol in membranes. *Mol Nutr Food Res*, 54, 641-651, 2010.

Barenholz, Y. Sphingomyelin-lecithin balance in membranes: composition, structure, and function relationships. In: Shinitzky, M., ed. Physiology of membrane fluidity. CRC Press, Boca Raton, FL, 131-173, 1984.

Buettner, GR. The pecking order of free radicals and antioxidants: lipid peroxidation, alpha-tocopherol, and ascorbate. *Arch Biochem Biophys*, 300, 535-543, 1993.

Catalá, A. An overview of lipid peroxidation with emphasis in outer segments of photoreceptors and the chemiluminescence assay. *Int J Biochem Cell Biol*, 38, 1482-1495, 2006.

Dietrich, C; Yang, B; Fujiwara, T; Kusumi, A; Jacobson, K. Relationship of lipid rafts to transient confinement zones detected by single particle *tracking Biophys J*, 82, 274-284, 2002.

Edidin, M. The state of lipid rafts: from model membranes to cells, *Annu Rev Biophys Biomol Struct*, 32, 257-283, 2003.

Gidwani, A; Holowka, D; Baird, B. Fluorescence anisotropy measurements of lipid order in plasma membranes and lipid rafts from RBL-2H3 mast cells. *Biochemistry*, 40, 1242.2-12429. 2001.

He, HT; Lellouch, A; Marguet, D. Lipid rafts and the initiation of T cell receptor signalling. *Semin Immunol*, 17, 23-33, 2005.

Helms, JB; Zurzolo, C. Lipids as targeting signals: lipid rafts and intracellular trafficking, *Traffic*, 5, 247-254, 2004.

Hess, ST; Kumar, M; Verma, A; Farrington, J; Kenworthy, A; Zimmerberg, J. Quantitative electron microscopy and fluorescence spectroscopy of the membrane distribution of influenza hemagglutinin. *J Cell Biol*, 169, 965–976, 2005.

Jacobson, K; Sheets, ED; Simson, R. Revisiting the fluid mosaic model of membranes. *Science*, 268, 1441-1442, 1995.

Kusumi, A; Suzuki, K. Toward understanding the dynamics of membrane-raft-based molecular interactions. *Biochim Biophys Acta*, 1746, 234-251, 2005.

Lange, Y; Swaisgood, MH; Ramos, BV; Steck, TL. Plasma membranes contain half the phospholipid and 90% of the cholesterol and sphingomyelin in cultured human fibroblasts. *J Biol Chem*, 264, 3786-3793, 1989.

Mueller, NS; Wedlich-Söldner, R; Spira, F. From mosaic to patchwork: matching lipids and proteins in membrane organization. *Mol Membr Biol*, 29, 186-196, 2012.

Nigam, S; Schewe, T. Phospholipase A2s and lipid peroxidation. Biochim. *Biophys. Acta*, 1488, 167-18, 2000.

Parton, RG; Richards, AA. Lipid rafts and caveolae as portals for endocytosis: new insights and common mechanisms. *Traffic*, 4, 724-738, 2003.

Pike, LJ. Lipid rafts: bringing order to chaos. *J Lipid Res*, 44, 655-667, 2003.

Plowman, SJ; Muncke, C; Parton, RG; Hancock, JF. H-ras, K-ras, and inner plasma membrane raft proteins operate in nanoclusters with differential dependence on the actin cytoskeleton. *Proc Natl Acad Sci USA*, 102, 15500-15505, 2005.

Pralle, AP; Keller, EL; Florin, K; Simons, K; Horber, JK. Sphingolipid-cholesterol rafts diffuse as small entities in the plasma membrane of mammalian cells. *J Cell Biol*, 148, 997-1008, 2000.

Shaikh, SR; Kinnun, JJ; Leng, X; Williams, JA; Wassall, SR. How polyunsaturated fatty acids modify molecular organization in membranes: insight from NMR studies of model systems. *Biochim Biophys Acta*, 1848, 211-219, 2015.

Sharma, P; Varma, R; Sarasij, RC; Ira, Gousset, K; Krishnamoorthy, G; Rao, M; Mayor, S. Nanoscale organization of multiple GPI-anchored proteins in living cell membranes. *Cell*, 116, 577-589, 2004.

Simons, K; Toomre, D. Lipid rafts and signal transduction, *Nat Rev Mol Cell Biol*, 1, 31-39, 2000.

Simons, K; Vaz, WL. Model systems, lipid rafts, and cell membranes, *Annu Rev Biophys Biomol Struct*, 33, 269-295, 2004.

Singer, SJ; Nicolson, GL. The fluid mosaic model of the structure of cell membranes. *Science*, 175, 720-731, 1972.

Slotte, JP. Sphingomyelin-cholesterol interactions in biological and model membranes. *Chem Phys Lipids*, 102, 13-27, 1999.

Vaz, WL; Almeida, PF. Phase topology and percolation in multiphase bilayers: Is the biological membrane a domain mosaic? *Curr Opin Cell Biol*, 3, 482-488, 1993.

Young, RM; Zheng, X; Holowka, D; Baird, B. Reconstitution of regulated phosphorylation of FcepsilonRI by a lipid raft-excluded protein-tyrosine phosphatise. *J Biol Chem*, 280, 1230-1235, 2005.

EDITOR CONTACT INFORMATION

Dr. Angel Catalá
Facultad de Ciencias Exactas
Universidad Nacional de La Plata
Casilla de Correo 16, Sucursal 4, (1900)
La Plata, Argentina
Email: catala@inifta.unlp.edu.ar

Index

L

M

N

O

P

R

S

T

V

Z